T0205614

Studies in Big Data

Volume 75

Series Editor

Janusz Kacprzyk, Polish Academy of Sciences, Warsaw, Poland

The series "Studies in Big Data" (SBD) publishes new developments and advances in the various areas of Big Data- quickly and with a high quality. The intent is to cover the theory, research, development, and applications of Big Data, as embedded in the fields of engineering, computer science, physics, economics and life sciences. The books of the series refer to the analysis and understanding of large, complex, and/or distributed data sets generated from recent digital sources coming from sensors or other physical instruments as well as simulations, crowd sourcing, social networks or other internet transactions, such as emails or video click streams and other. The series contains monographs, lecture notes and edited volumes in Big Data spanning the areas of computational intelligence including neural networks, evolutionary computation, soft computing, fuzzy systems, as well as artificial intelligence, data mining, modern statistics and Operations research, as well as self-organizing systems. Of particular value to both the contributors and the readership are the short publication timeframe and the world-wide distribution, which enable both wide and rapid dissemination of research output.

Indexed by zbMATH. All books published in the series are submitted for consideration in Web of Science.

More information about this series at http://www.springer.com/series/11970

Anna Felkner · Youki Kadobayashi ·
Marek Janiszewski · Stefano Fantin ·
Jose Francisco Ruiz · Adam Kozakiewicz ·
Gregory Blanc

Cybersecurity Research Analysis Report for Europe and Japan

Cybersecurity and Privacy Dialogue Between Europe and Japan

Springer

Anna Felkner🆔
NASK—Research and Academic
Computer Network
Warsaw, Poland

Marek Janiszewski🆔
NASK—Research and Academic
Computer Network
Warsaw, Poland

Jose Francisco Ruiz
Atos (Spain)
Madrid, Spain

Gregory Blanc🆔
Institut Polytechnique de Paris
Institut Mines-Télécom
Evry-Courcouronnes, France

Youki Kadobayashi
Nara Institute of Science
and Technology
Ikoma, Japan

Stefano Fantin
Center for IT and IP Law
KU Leuven
Leuven, Belgium

Adam Kozakiewicz🆔
NASK—Research and Academic
Computer Network
Warsaw, Poland

ISSN 2197-6503   ISSN 2197-6511   (electronic)
Studies in Big Data
ISBN 978-3-030-62314-2   ISBN 978-3-030-62312-8   (eBook)
https://doi.org/10.1007/978-3-030-62312-8

This Springer imprint is published by the registered company Springer Nature Switzerland AG
The registered company address is: Gewerbestrasse 11, 6330 Cham, Switzerland

Foreword

At a time where certain countries have a tendency to look inwards, this book offers a perfect example of the benefits for countries to, instead, look outwards when it comes to international collaboration to address cybersecurity issues.

It contains a massive amount of information collected by European and Japanese experts, sitting and working together for several years to produce this high quality material. They offer several comparative surveys to help understand the various efforts taking place in Europe and in Japan. The legal field is covered, as well as funding mechanisms and cyber strategies in a number of countries. It also highlights the impressive amount of effort invested in a number of different projects over the years.

The authors also take the time to assess the differences, the strengths and weaknesses that exist both in the European and Japanese scenes. By doing so, they highlight areas for future joint beneficial collaborations.

Last but not least, the book proposes an interesting survey of the various business and industrial associations, as well as standard organizations, present in both places and the role they can play, nationally or internationally.

This book is a must read for whoever is active in the cybersecurity area. Its broad spectrum will render it appealing to a large audience but its depth is such that even experts in specific fields will very likely learn something new in the chapters dedicated to their own domains.

The authors, editors and contributors have to be commended for this impressive publication.

February 2020

Marc Dacier
Digital Security Department Head
EURECOM
Sophia Antipolis, France

Preface

This book contains the most important findings related to cybersecurity research analysis for Europe and Japan, which were collected during the EUNITY project. The EUNITY project ran from 1st June 2017 to 31st May 2019, and aimed to encourage, facilitate and develop dialogue between Europe and Japan, on cybersecurity and privacy research and innovation trends and challenges, in order to foster and promote cybersecurity activities in both regions. One of the main objectives of the EUNITY was identification of potential opportunities for future cooperation between European and Japanese R&I ecosystems and policymakers by identifying and mapping relevant legislation, policies and cybersecurity agendas, roadmaps and timelines at the EU level (NIS platform, cybersecurity cPPP) and at the National level, as well as in Japan, clearly identifying and prioritizing the joint areas of interests and ensuring that the collected information is made widely available through modern means of communication. This publication strictly corresponds to that objective; however, it is not part of the project.

We hope that this publication would be a valuable position to foster cybersecurity research in both regions and to facilitate further dissemination of the results obtained.

Warsaw, Poland
December 2019

Anna Felkner
Marek Janiszewski

Acknowledgements

The authors would like to thank many other researchers who participated in the EUNITY project. Special thanks should be attributed to:

- Hervé Debar
- Christophe Kiennert
- Kazuya Okada
- Despoina Antonakaki
- Christos Papachristos
- Sotiris Ioannidis
- Pedro Soria-Rodriguez
- Dmitriy Pap
- Paweł Pawliński

and many others participants of the three EUNITY workshops (which took place in Tokyo, Brussels and Kyoto in 2017 and 2019).

The authors also want to thank the European Commission for financing the project, which has facilitated further activities in this area, such as this publication.

Contents

Acronyms

AI	Artificial Intelligence
AIS	Automated Indicator Sharing
APPI	Act on the Protection of Personal Information
APT	Advanced Persistent Threat
ASEAN	Association of Southeast Asian Nations
CAO	Cabinet Office
CEF	Connecting Europe Facility
CERT	Computer Emergency Response Team
CFSP	Common Foreign and Security Policy
CIA	Confidentiality, Integrity and Availability
CII	Critical Information Infrastructure
CIIP	Critical Information Infrastructure Protection
cPPP	Contractual Public-Private Partnership
CSDP	Common Security and Defence Policy
CSIRT	Computer Security Incident Response Team
CSP	Cyber Security Package
CSTI	Council for Science, Technology and Innovation
DDoS	Distributed Denial of Service
DHS	Department of Homeland Security
DPA	Data Protection Authority
DPIA	Data Protection Impact Assessment
DSM	Digital Single Market
DSP	Digital Service Providers
EC	European Commission
ECSO	European Cyber Security Organisation
EDPB	European Data Protection Board
ENISA	European Union Agency for Network and Information Security
EOS	European Organisation for Security
ERA	European Research Area
ETSI	European Telecommunications Standards Institute

FP7	The seventh Framework Programme
GDP	Gross Domestic Product
GDPR	General Data Protection Regulation
H2020	Horizon 2020
HPC	High Performance Computing
ICT	Information and Communication Technologies
IoT	Internet of Things
IPA	Information-technology Promotion Agency
ISAC	Information Sharing and Analysis Center
JSPS	Japan Society for the Promotion of Science
LEA	Law Enforcement Agency
MERRA	Ministry of Education, the Research and Religious Affairs
METI	Ministry of Economy, Trade and Industry
MEXT	Ministry of Education, Culture, Sports, Science and Technology
MIC	Ministry of Internal Affairs and Communications
ML	Machine Learning
MS	Member State
NCBR	National Centre for Research and Development
NCSS	National Cyber Security Strategy
NGN	Next Generation Network
NICT	National Institute of Information and Communications Technology
NIS	Network and Information Systems Directive
NISC	National center of Incident readiness and Strategy for Cybersecurity
NISC	National Information Security Center
NIST	National Institute of Standards and Technology
NSRF	National Strategic Reference Framework
OES	Operators of Essential Services
PET	Privacy Enhancing Techniques
PIP	Public Investment Programme
PIPC	Personal Information Protection Commission
R&D	Research and Development
R&D&I	Research, Development and Innovation
R&I	Research and Innovation
SCOPE	Strategic Information and Communications R&D Promotion Programme
SIP	Strategic Innovation Promotion Program
SME	Small and medium-sized enterprise
SRIA	Strategic Research and Innovation Agenda
TRL	Technology Readiness Level
UK	United Kingdom
US	United States

Chapter 1
Introduction

The goal of this book is to establish a clear picture on the cybersecurity and privacy domain in both regions by analyzing existing roadmaps. First, it will identify the appropriate methodology applicable to the information collected, and then analyse cybersecurity priorities in both the EU and Japan, in order to produce a background document on the status and priorities of cybersecurity and privacy research and innovation activities in Europe and Japan.

The scope of the book includes a review of the mechanisms used to finance research and innovation in Europe (both European and national funds, including non-EU level international cooperation) and Japan. It also includes an overview of the main research directions in the field, identifying the strong and weak points in the European and Japanese research landscape, looking for topics of common interest, where cooperation opportunity is clear, as well as topics where some aspects are covered asymmetrically, allowing even greater synergy.

The book also shows the analysis of the current role and activities of various entities (SMEs, research institutions, CSIRTs, LEAs, etc.) in research and innovation in Europe and Japan, looking for possible asymmetries that increase the value of possible cooperation. An analysis of long-term research programmes at national and international levels has also been carried out. The main objective of this analysis is to find thematic parallels between the EU and Japan that may create opportunities for either co-financing of joint EU-Japan projects or at least synchronization of efforts enabling cooperation between topically linked European and Japanese projects. The purpose of the analysis in the book is only to indicate the most visible similarities and differences.

The book was created on the basis of the two main reports created during EUNITY project, namely: "D3.1 Preliminary version of the Cybersecurity Research Analysis Report for the two regions" [1] and "D3.2 Revised version of the Cybersecurity Research Analysis Report for the two regions" [2]. The above mentioned two reports have been used also to create a thorough analysis and a set of conclusions, which can be found in the following reports: "D4.1 Description of gaps and future challenges" [3] and "D4.2 Strategic research and innovation agenda" [4].

© The Author(s), under exclusive license to Springer Nature Switzerland AG 2021
A. Felkner et al., *Cybersecurity Research Analysis Report for Europe and Japan*,
Studies in Big Data 75, https://doi.org/10.1007/978-3-030-62312-8_1

Elicitation of common topics of interest will establish a clear picture on the cybersecurity domain in both regions by analyzing existing cybersecurity and privacy roadmaps.

Chapter 2 (*Legal and Policy Aspects*) focuses on the analysis of regulatory documents, EU directives as well as laws and legal frameworks. It elicits the common regulatory aspects of cybersecurity and privacy, for example obligations for monitoring, certification, protection of personal data and information exchange. This mainly applies to policy-makers and CERT communities, although the business community is certainly impacted.

Chapter 3 (*Research and Innovation Aspects*) focuses on the analysis of research agendas, programs, project calls and roadmapping projects as well as on showing the available mechanisms to finance cybersecurity research.

Chapter 4 (*Industry and Standardization Aspects*) focuses on the analysis of the output of clusters and industry associations. It also includes national strategic efforts to create or develop business strategies. With regard to standards, it covers the entire range of standardization bodies.

References

1. Felkner, A. Janiszewski, M. (eds): Preliminary version of the Cybersecurity Research Analysis Report for the two regions (2018). https://www.eunity-project.eu/m/filer_public/53/4a/534abeb6-6532-4c59-a4ae-22ac91b3d885/eunity-d31.pdf. Cited 15 Nov 2019
2. Felkner, A. Janiszewski, M. (eds): Revised Version of the Cybersecurity Research Analysis Report for the Two Regions (2019). https://www.eunity-project.eu/m/filer_public/6d/07/6d078aaa-fba9-4101-8119-951041e67dbb/eunity-d32.pdf. Cited 15 Nov 2019
3. EUNITY project. Description of gaps and future challenges (2018). https://www.eunity-project.eu/m/filer_public/98/f5/98f575cd-6160-4ea0-bc37-aa1112288be1/eunity-d41_v4.pdf. Cited 15 Nov 2019
4. EUNITY project. Strategic research and innovation agenda (2019). https://www.eunity-project.eu/m/filer_public/d1/d8/d1d89adf-8a92-4633-9b95-3863a5d8f96c/eunity_d42.pdf. Cited 15 Nov 2019

Chapter 2
Legal and Policy Aspects

The present chapter elicits the legal and regulatory landscape in the fields of privacy and cybersecurity in the European Union and Japan. With respect to the former, attention is paid to the substantial reform on privacy and data protection brought forward mostly by the so-called GDPR (General Data Protection Regulation), while the cybersecurity analysis focuses on the cybersecurity strategy of the European Union and the NIS (Network and Information Security) Directive.

With regard to the Japanese landscape, privacy and cybersecurity is analysed through the lens of the Act on Personal Information and the Basic Act on Cybersecurity, respectively. Inter alia, such an analysis tries to draw parallels and differences with the above-mentioned European acts, so as to offer a comparative overview of the main legal concepts therein and of the possibilities created by both legislations to engage in cooperative efforts and information sharing.

As this chapter outlines, gaps and differences still persist amongst the two regions, in spite of the many similarities and overlaps. For instance, the approaches taken by the two Acts on cybersecurity delineate a fundamental gap between the two. Whilst the scope of the European NIS Directive applies to organizations and legal persons, the Japanese Basic Act includes a significant section dealing with the obligations on the security of the information systems that *individuals* should comply with. Similarly, the two laws on data protection have different approaches, namely within the sanctionatory regime (much stricter, in principle, in the GDPR), and in the inclusion within the definition of personal information of online identifiers such as IP addresses and alike (which have a wider protection in the European framework).

However, it must be noted that the legislative process still needs to be interpreted by the two jurisprudences, and a new EU Act (the Cybersecurity Act) has been issued recently. Such a legal instrument, if taken and approached correctly, might actually decrease the legal gap between the two regions, facilitating the cooperation and the information sharing, towards a more secure and sustainable future by ways of a brand new permanent mandate to the EU Cybersecurity Agency (ENISA) and a set of cybersecurity certification frameworks. This is why the present chapter aims

A. Felkner et al., *Cybersecurity Research Analysis Report for Europe and Japan*, Studies in Big Data 75, https://doi.org/10.1007/978-3-030-62312-8_2

at outlining a preliminary and high level legal and policy analysis on commonalities and differences between EU and Japan in privacy and cybersecurity laws.

2.1   The European Landscape

The present section aims at outlining the current and imminent landscape of EU law and policy as far as privacy, data protection and cybersecurity are concerned. The recent years were characterized by an intense effort by the European legislator aimed at reforming and modernizing privacy and cybersecurity laws, standards and practices across Europe.

   The rise of modern technologies and new forms of human and digital interactions has brought into light the weaknesses of the existing legislation, which turned to be perceived by many as fragmented [1] and outdated. This "obsolete" state of play was accompanied by the increasing number of cyber threats that European stakeholders are facing [2] in both public and private sectors: growing security and privacy risks have become a landmark blocker for the achievement of a competitive and secure digital single market.

   Against this backdrop, the above mentioned scenario triggered vibrant societal discussions [3] which, ultimately, brought the political arena to initiate a long decision making process, leading to a substantial reform of the legislative and policy landscape in both data protection and cybersecurity fields [4]. This section will therefore seek to outline the main legal and policy efforts by the European Union addressing such discrepancies. The first chapter will deal with the General Data Protection Regulation (GDPR) [5] and the overall privacy reform package, while the second will focus on the legislative initiatives in terms of cybersecurity, namely the Network and Information Security Directive (NIS) [6] and the EU Commission's European Cyber Security Strategy.

   The analysis that will follow aims at supporting both European and Japanese partners (policy and decision makers, CERT representatives and businesses) in further understanding what major implications for their business processes arise from the said evolving legal landscape. The objective of outlining such elements shall well serve to bring closer the two regions, thus facilitating a reciprocal understanding of opportunities and boundaries inherent in the legal landscapes under scrutiny below.

2.1.1   Privacy and Data Protection

2.1.1.1   The General Data Protection Regulation

On May 25th, 2018, the General Data Protection Regulation (GDPR) entered into force, repealing the existing Directive 46/95/EC. It marked a considerable change for the European privacy landscape, inter alia because of the legal instrument chosen to convey such novelties. As a Regulation, the GDPR has in fact binding legal force across all Member States of the European Union. Conversely, the Directive 46/95/EC [7] rolled out a set of provisions that had to be implemented within each domestic legal framework. It goes without saying that the GDPR aims therefore at a higher level of harmonization than the one previously in force, with less room for domestic interpretation and more legal certainty across the European territories [9].

Technology Neutral. One crucial aspect of the GDPR is enshrined in Recital 15, stating that "in order to prevent creating a serious risk of circumvention, the protection of natural persons should be technologically neutral and should not depend on the techniques used". Firstly introduced in the EU law landscape in 2002 within the so-called e-Privacy Directive [8], for the sake of this analysis the meaning of this concept is that the principles enshrined in the Regulation are drafted and shall apply regardless of the technological advancements, and that the same regulatory framework shall not influence innovation towards an intended outcome [10]. Technological neutrality should therefore ensure an endurable application of the GDPR in spite of short and mid-term technological innovations.

Scope and Grounds for Processing

(Extra-territorial) Scope. One of the main changes from the existing legislation to the GDPR refers to the scope of application of the GDPR. Its provisions in fact apply with a broad interpretation of the principle of territoriality: Article 3 prescribes that the Regulation applies to personal data processing by those controllers and processors[1] (a) established in the European Union, irrespective of where the processing activity takes place, or (b) process personal data of European citizens, irrespective of their place of establishment. Such broad rules aim at a wider reach, actually including non-European businesses and organizations processing data of EU subjects, including the activities of behavioral monitoring and provisions of goods and services (including free of charge).

Grounds for Processing. Article 6(1) provides for six legal basis that justify the processing of personal data by controllers, namely[2]:

(a) the data subject has given **consent** to the processing of his or her personal data for one or more specific purposes;
(b) processing is necessary for the **performance of a contract** to which the data subject is party or in order to take steps at the request of the data subject prior to entering into a contract;
(c) processing is **necessary** for compliance with a **legal obligation** to which the controller is subject;

---

[1]See GDPR [5], Article 4.7 and 4.8: (7) 'controller' means the natural or legal person, public authority, agency or other body which, alone or jointly with others, determines the purposes and means of the processing of personal data; where the purposes and means of such processing are determined by Union or Member State law, the controller or the specific criteria for its nomination may be provided for by Union or Member State law; (8) 'processor' means a natural or legal person, public authority, agency or other body which processes personal data on behalf of the controller; See also: [9].
[2]GDPR [5], Article 6(1). Emphasis added by the authors.

(d) processing is **necessary** in order to protect the **vital interests** of the data subject or of another natural person;

(e) processing is **necessary** for the performance of a **task carried out in the public interest** or in the **exercise of official authority** vested in the controller;

(f) processing is **necessary** for the purposes of the **legitimate interests pursued** by the controller or by a third party, **except** where such interests are overridden by the interests **or** fundamental rights and freedoms of the data subject which require protection of personal data, in particular where the data subject is a child.

Data Subjects' and Information Rights

The GDPR has broadened the scope of the rights of the data subjects towards those controllers responsible for processing their personal information [11]. As a result, the Regulation aims at providing full control over personal data to European citizens [12], thus protecting them against potential interferences or violations of their fundamental right to data protection [13]. From the controller perspective, such data subjects' empowerment leads to an increment of the obligations they need to observe and respect.

Table 2.1 provides a summary of the data subject's rights as enshrined by GDPR.

The increment in citizens' control over their personal data also comes from the right to an effective remedy (Article 79). It adds up to the abovementioned rights as a complementary one, and seeks to ensure citizens to have proper access to regulatory justice by means of the right to lodge a complaint to the competent supervisory authority, pursuant to Article 77, as well as to the ordinary justice.

Obligations on Certification and Codes of Conduct

Pursuant to the principle of accountability,[3] the General Data Protection Regulation provides for a set of articles encouraging the establishment of codes of conduct and certification schemes. Additionally to the assessment of the Regulation, for the sake of this analysis, it is useful to look at ENISA's Report [14] issued in November 2017, where certification is considered as a "conformity assessment activity" [14] performed by a third party (as opposed to a first-party self-assessment), including a set of rules and procedures ("scheme") to perform such an exercise, resulting in a "certificate" or "seal" validating it. With a view to the future, a number of ENISA's policy papers[4] address certification as a major component of the next regulatory landscape on privacy and cybersecurity.

Article 42 of the GDPR asks Member States, the European Commission, the EDPB[5] to encourage, at the Union level, the creation of certification schemes as a way to ensure and promote compliance with the Regulation, as well as to disseminate good practice on privacy and data protection across all sectors within and out of European territories, particularly for SMEs and businesses alike. Among others, GDPR aims at ensuring through such certification schemes the existence of appropriate data

---

[3] Article 24, GDPR [5].

[4] See for instance [15].

[5] The European Data Protection Board.

Table 2.1 Summary of the data subject's rights as enshrined by GDPR

Article	Right	Brief description
12,13 and 14	Right to information	In light of an improved transparency of data processing, data subjects have the right to request (free of charge) for what purposes and by whom exactly (third parties included) their data are processed
15	Right to access	The request may also pertain a copy of such data (see above), that must be provided electronically, in an easily-readable format and without undue delay
16	Right to rectification	If personal data is inaccurate or incomplete, the data subject has the right to ask for them to be rectified
17	Right to erasure (Right to "be forgotten")	Data subjects have the right (under qualified circumstances) for their data to be erased from the controllers' dataset.[a]
18	Right to restriction of processing	Controllers are bound by the right of the data subject to ask for their personal information to be suspended from the processing
20	Right to data portability	Data subjects have the right for their data to be transferred from one controller to another (or directly to the data subjects themselves) in an automatic and easily readable format
21	Right to object	Right to contest the processing of personal data by controllers
22	Right not to be subject to automated decisions	Data subjects have the right not to be subject to decisions based on solely automated means (including profiling), which leads to legal effects for the individuals

[a]C-131/12—Google Spain and Google: Judgment of the Court (Grand Chamber), 13 May 2014 Google Spain SL and Google Inc. v Agencia Española de Protección de Datos (AEPD) and Mario Costeja González—Request for a preliminary ruling from the Audiencia Nacional

protection safeguards in the case of **personal data transfers** to third countries or international organizations. Against this backdrop, the Regulation provides for a set of requirements to be taken into account by law makers and regulatory bodies when establishing such certification mechanisms, such as:

- The voluntariness of the certification for controllers and processors;
- A certification criteria to be set by EDPB;
- The maximum length (renewable) of certification shall be of three years;
- A publicly available register of all the "approved" certification mechanisms made publicly available by EDPB.

It is noteworthy to point out that the adherence to a certification scheme does not exclude or impact the potential liability of controllers or processors that are found in violation of the GDPR, nor does it exclude capacities and responsibilities of data

Table 2.2 Three different avenues to create certification bodies

Avenue A	Avenue B	Avenue C
Accreditation by a Data Protection Authority (or the European Data Protection Board, in the case of the European Data Protection Seal ex Article 70 GDPR)	Accreditation by the National Accreditation Body on the basis of the Accreditation Regulation and the ISO/IEC 17065:2012 standard and additional requirements in the field of data protection provided by the Data Protection Authority	Both authorities, namely the National Accreditation Body and the competent Data Protection Authority, collaborating in this task

protection authorities across Europe with regard to their enforcement and oversight powers.

With reference to third parties, as ENISA remarks in its report, Article 43 "explicitly requires accreditation for the certification body" [14], thus pursuant to the rules laid down in Regulation 765/2008 (Accreditation Regulation) [16]. Such bodies, which shall follow the principles of independence, reliability and competence [14] may thus be created in the context of three different avenues, as unfolded in Table 2.2.

Following the assessment of the GDPR, ENISA identifies a number of challenges and open questions with reference to certification schemes as foreseen by the Regulation, followed by a number of recommendations aimed at mitigating such challenges. In particular, ENISA stresses that national DPAs and national certification bodies shall adopt a common and coordinated approach on "inception and deployment of GDPR data protection certification mechanisms" [14], under the coordination of the European Commission and of the EDPB. This will help avoiding the proliferation of certification mechanisms with substantially and formally diverse angles and potentially conflicting approaches towards the same subject matters. Furthermore, it is encouraged that the same players shall work together to "stimulate the exchange of best practices and lessons learnt from certification practices in other well established domains (e.g.. cybersecurity)" [14].

As a complementary note, GDPR foresees in Article 40 the creation of sector-specific **codes of conduct** "intended to contribute to the proper application of this Regulation", with particular reference to SMEs. At the time of writing, a number of industry organizations, business associations and various sectorial networks are in the process of starting to set up the basis for a sectorial code of conduct pursuant to the improvement of GDPR compliance by all actors (some of them are being amended, as already in place under the current Directive). Such codes of conduct will then need to be revised and approved by competent data protection authorities and the EDPB (in case the Code applies to organizations established in more than one Member State). The main points of the codes of conduct mentioned by Article 40 shall address the following elements of GDPR: fair and transparent processing, grounds for legitimate interests of the controller, data collection activities, pseudonymization, information rights, protection of minors' personal data, security, breach notification, data transfers and forms of alternative dispute resolutions. Interestingly, Paragraph

3 of Article 40 extends the possibility to adhere to such codes also to players not having to abide by the GDPR, for example "by controllers or processors that are not subject to this Regulation pursuant to Article 3 in order to provide appropriate safeguards within the framework of personal data transfers to **third countries** or international organizations". It goes without saying that such clause seeks to enable GDPR-compliant relationships with non-EU countries, exporting European privacy standards as a baseline for the engagement of external partners.

Data Security: Monitoring Obligations and Breach Notifications

One of the points of contact between privacy, data protection and cybersecurity in the GDPR is certainly represented by the set of provisions regarding data security. Obligations, such as the duty to undertake monitoring activities identified in data protection impact assessments (DPIAs), provisions related to technical and organizational measures and breach notification obligations present elements of overlap between the two disciplines which will be further articulated below.

Monitoring Obligations. One of the most important elements to highlight relates with the shift that the legislation imposes on controllers with regard to the assessment on whether a data processing operation results in a risk for individuals' rights and freedoms, i.e. from a legally-based approach to a more risk-based approach. Data protection impact assessments thus become a mandatory obligation within the GDPR (Article 35) [5] auspices under a certain number of conditions. It is fair to point out that impact assessments were mentioned already in the existing Directive, but the move towards more compulsory provisions supports the argument that, from the legislator viewpoint, data protection and information security compliance are getting close as never before: the rationale behind this choice is to bring controllers to reason about privacy according to a priority-risk mindset, an approach already in use in many cyber-security ecosystems.

Coming down to the Regulation, Article 35 demands data controllers to undertake DPIAs anytime they deem a data processing operation results in a "high risk to the rights and freedoms of natural persons" [5]. A number of interpretative issues arise from the text of the Regulation, hereby briefly outlined.

With regard to the assessment of the so-called high risk, it should be noted that WP29[6] provides further guidance for unlocking the concept [17]; additionally, for the sake of this analysis, it is useful to point that a number of processing operations are subtracted from the arbitrary scrutiny of the data controllers, as they are considered not to necessitate a DPIA by default. As Article 35(5) states in fact, data protection authorities may issue a public list of processing operations for which such exercise would not be required.

WP29 provides for a set of non-exhaustive operations to be evaluated as potentially risky and therefore falling under the DPIA requirement. The list includes:

- evaluation and scoring,
- automated-decision making with legal or similar significant effect,

---

[6]Article 29 Data Protection Working Party, the panel composed by members of EU Member States' privacy regulators, replaced by the EDPB under the GDPR regime.

- systematic monitoring,
- involvement of sensitive data,
- data processed on a large scale,
- datasets that have been matched or combined,
- data concerning vulnerable data subjects,
- innovative use or application of technological or organizational solutions,
- data transfer across borders outside the European Union,
- if the processing "prevents data subjects from exercising a right or using a service or a contract" [17].

Security of Processing. It is important to note that GDPR promotes the use of Privacy Enhancing Techniques (PETs) as a way to achieve compliance and security in a given organization. For instance, tools and methods like encryption or pseudonymization are explicitly mentioned in the text of the law as a best practice.

However, GDPR expects data controllers to assess the data processing operation before it starts and, according to the potential risks for the security of processing itself, to implement adequate technical and organizational measures which are deemed as appropriate for the business case in question.

Notifications for data breaches. Data breach notifications represent a novelty in comparison with the existing Directive 95/46: on October 2017, WP29 issued guidelines on breach notification [18], which will help us unlocking some of the basic concepts behind Article 33.

Article 4(12) GDPR defines personal data breaches as "a breach of security leading to the accidental or unlawful destruction, loss, alteration, unauthorized disclosure of, or access to, personal data transmitted, stored or otherwise processed". It goes without saying that the element of security here is strictly linked with the compromising of at least one of the three components of the so called CIA triad, which in information security refers to Confidentiality, Integrity and Availability [19].

The above mentioned Article 33 requires data controllers and processors to have 72 h from the moment they become aware (i.e., when there is a "reasonable degree of certainty that a security incident has occurred" [19]) of the breach to take actions towards the following stakeholders:

- Data protection authorities, unless there is the unlikelihood that the breach falls into the category of those incidents causing a risk for the freedoms of individuals.
- Data subjects, in the event that high risks for the freedoms of the individuals are likely to occur.

Interestingly, WP29 makes it clear that "an incident resulting in personal data being made unavailable for a period of time is a security breach (and should be documented), yet depending on the circumstances, it may or may not require notification to the supervisory authority and communication to affected individuals" [19]. This statement clarifies that controllers need to document a larger number of incidents than the ones for which they will need to initiate a notification to the DPA, thus applying Article 33(5) which is clearly inspired by the principle of accountability. Furthermore, WP29 specifies that even incidents causing the temporary unavailability of personal data fall under the category of incidents ex Article 33(5).

Furthermore, it is fair to assume that the reporting duty does not imply a one-end line of communication only. The aim of laying down personal data breach reporting in the GDPR is to enable a trustworthy relationship between controllers and competent data protection authorities, where the latter are expected to provide assistance in return to a notifying controller, possibly guiding them towards a situation of low-risk. Mutual updates and follow-ups are mentioned and welcome by WP29 [19], as a demonstration of the need for enhancing and facilitating smooth cooperation between regulators and controllers.

Another important point for data security is the obligation of embedding privacy by design in both development and deployment of a system processing personal information [5]. Concepts of information security like data minimization, access control, access rights, activity logs, are all functional for meeting such requirement, and ultimately offer a paradigm of how the GDPR is actually influenced by features, elements and notions that are typical of a security context.

2.1.2 Cybersecurity

The EU Cyber Security Strategy in the Digital Single Market Context. The Digital Single Market (DSM) is an initiative launched in 2015 by the European Commission with the aim of fostering the role of the EU as a global leader in the digital economy [20]. The DSM has three major policy areas:

- Providing better access for consumers and business to online goods
- Creating the right environment and conditions for digital networks and services (telco reform, cybersecurity cooperation, audiovisual reform)
- Ensuring that economy, business and society are positively impacted by digitization.

Developing stronger cybersecurity rules constitutes a prominent objective in such context, as ensuring network security has a strong link with the growth of Europe's competitiveness in the digital economy. To this end, within the mid-term review[7] of the DSM initiative, a new Cyber Security Package (CSP) was presented, building on the previous 2013 EU Cyber Security Strategy. Amongst other initiatives, it is noteworthy to underline the following actions foreseen in the CSP, particularly aimed at fostering European cyber resilience:

1. **The European Network and Information Security Agency (ENISA)**. The role of ENISA, established in 2004 [22], is intended to expand and become the so-called "EU Cybersecurity Agency" [23]. The ambitious proposal of the Commission is in fact to turn the agency into a centralized body that provides constant support to Member States and EU Institutions in key areas. Such initiative is rolled in a number of policy actions, including but not limited to the recently issued "Cybersecurity Act" [24].

---

[7]The Digital Economy and Society Index (DESI) 2017 (European Commission).

2. **The Network and Information System Directive (NIS)**. After having adopted it in April 2016, the European Union aims at a swift implementation across its Member States. Laying down rules for application of the NIS Directive should serve the scope of harmonizing a common approach towards implementation and interpretation of the main provision of the abovementioned Directive [25] . About the core principles of the NIS, more will be articulated in the next sections.
3. **Cyber Defense**. Under the Common Foreign and Security Policy (CFSP) [26], the European Union recently adopted a framework for a joint EU diplomatic response to malicious cyber activities (the "cyber diplomacy toolbox") aims at harmonizing common responses to external cyber threats against the Union and its Member States [27].

2.1.2.1 The NIS Directive

As briefly mentioned above, the NIS Directive is the other crucial piece of EU legislation that is set to uplift the cybersecurity landscape alongside the GDPR. Differently from the latter, the NIS has no direct applicability within Member States (and does not strictly pertain to the protection of personal data, as we will see). This means that EU countries had an obligation to transpose the framework within a fixed amount of time (expired in April 2018), with the consequence of resulting in twenty-eight potentially different interpretations of the same provisions. However, despite the legislators' choice to adopt a minimum harmonization approach [6], the Directive has been long awaited and is supposed to lay down, along with its domestic transposing acts, those minimum requirements to ensure an acceptable and competitive level of cybersecurity, awareness and resilience amongst the Union [28]. Specifically, it does roll out three goals to achieve:

1. Increasing EU-wide cybersecurity capabilities;
2. Enhancing multi-actors cooperation;
3. Laying down security measures and incident reporting obligations for certain players.

The scope

The first category of stakeholders that are subject to the NIS Directive is the so-called "Operators of Essential Services" (OES), defined in Article 4 as any private or public entity that cumulatively meets two requirements:

1. Falls under article 5(2), i.e. provides services depending on network and information systems which are essential for the maintenance of critical societal and/or economic activities and for which a potential incident would cause disruptive effects[8] on the provision of that same service.

---

[8]Parameters and criteria for determining "disruptive effects" are further established in Article 6.

2. Is of a type referred to in Annex II[9] of the same NIS Directive.

The second category refers to "Digital Service Providers" (DSP), again defined as those services meeting the following two requirements:

1. Delivering digital services as enshrined in Article 1(1) of Directive 1535/2015.[10]
2. The service is an online marketplace, an online search engine or a cloud computing service [6].

Two things are worth being noted on the scope of the NIS Directive. Firstly, a clause in the NIS provides for an exception for the referenced operators to be exempted from abiding to the Directive. It is in fact the case of players operating in particular areas which are highly sector-regulated and thus already meet in principle high security standards due to their compliance with other laws. Secondly, the Directive does not only address the categories mentioned above. Indeed, it also requires Member States to establish CSIRTs (Computer Security Incident Response Teams), pursuing a more coordinated pan-European resilience to information security incidents, as well as competent NIS-responsible authorities (each domestic transposition law clarifies each Member State's designated body).

Security Measures and the Role of Standards

Similarly to what the GDPR requires, so does the NIS with respect to the security measures to be taken by OESs and DSPs separately. "Technical and organizational measures" are thus embedded in such security obligations laid down by Articles 14, 15 and 16. Beyond the mere legal analysis of such obligations, it is interesting to note the approach of the legislator when referring to the state of the art information security technology to be implemented by OESs and DSPs [28]. Article 19 in fact, encourages the "use of European or internationally accepted standards and specifications relevant to the security of network and information systems" as means to promote a convergent approach to Article 14 and 16 (respectively addressing OESs and DSPs).

As a way of regulatory support to the reference above, the second paragraph of Article 19 relies on the role of ENISA in coordinating the effort with Member States to roll out guidelines on the technical areas to be addressed and on the already existing technical standards to draw up from.

It needs to be highlighted here that compliance with a certain standard does not only permit an organization to demonstrate a certified and accountable conduct through the rules therein. The process of standards negotiations in fact (see below), does also enhance the cooperative efforts amongst various industry sectors, as well as horizontally throughout the supply chain [21]. It provides for a line of junction

---

[9]Annex II groups activities and businesses in seven macro-areas: energy, transport, banking, financial market infrastructures, health sector, drinking water supply and distribution, digital infrastructures.

[10]The definition includes the following description: "service" means any Information Society service, that is to say, any service normally provided for remuneration, at a distance, by electronic means and at the individual request of a recipient of services.

between organizations, sectors and geopolitical regions which is vital in the context of cybersecurity, as very often the response to a threat is put in place at the global scale.

Irrespective of the presence of a legal framework in a determined sector, it should be noted that standards may address certain issues which could not be directly solved by legislation otherwise, since they very often pertain to problems and constraints that are perceived by a number of sectors but are not enough widespread to be included in a more high-level legal text with a much wider scope.[11]

However, back in 2005, Weiss and Weber found more than 40 cybersecurity standards and put them in a non-exhaustive, paradigmatic list at margin of their paper [30]. Ever since, the number has increased and today it comes as almost impossible to map all those relevant standards applicable in the information security domain. It is clear that fragmentation has the advantage of being sufficiently specific to tackle narrow and particular issues, while leaving the law to provide for a more high-level approach with a robust framework. The challenge observed is ultimately, to understand what is the best approach in order to achieve a cooperative cybersecurity relationship between selected trade partners like the EU and Japan. It is now too soon to give a consistent evaluation, as the NIS Directive will soon be in force in all 28 Member States, and the GDPR could be considered as a *freshman* in the legal landscape. What we can say however, is that the role of standards has not decreased over time and in spite of the existence of regulations and laws, it still provides for a normative rulebook to many industry fields.

Processes

From a regulatory perspective, standards in Europe can either be seen as an alternative to "hard regulation" or an example of self-regulation. They can be either voluntary (market-driven), consensus-based or approved and recognized by a body (for instance, ETSI, NIST, ISO). According to the EU Regulation 1025/2012, *"standard" means a technical specification, adopted by a recognized standardization body, for repeated or continuous application, with which compliance is not compulsory, and which is one of the following:*

(a) *"international standard" means a standard adopted by an international standardization body;*
(b) *"European standard" means a standard adopted by a European standardization organization;*
(c) *"harmonized standard" means a European standard adopted on the basis of a request made by the Commission for the application of Union harmonization legislation;*
(d) *"national standard" means a standard adopted by a national standardization body* [31].

The Danish Standard Foundation enlisted the five-step process [32] with which a standard is developed:

---

[11] See also: [29].

(a) Proposal and approval phase
(b) Drafting and commenting phase
(c) Public consultation and vote
(d) Draft to formal vote
(e) Final Publication

Legal effects take place by way of three different avenues, i.e. by incorporation in contracts, implicit legal significance or reference in normative or statutory text [33].

Breach Notification and Incident Reporting

The NIS Directive (Article 16) requires Member States to ensure that digital service providers notify the competent domestic authority of any incident resulting in a substantial impact on the provision of a service without undue delay.

Firstly, the competent authority or CSIRT shall be provided by the breached organizations with all the information necessary to assess any cross-border impact of the incident. If an incident impacts more than one Member State, it will be responsibility of the first[12] Member State's authority to inform the second one about the breach. Secondly, it needs to be noted that the term "substantial impact" as expressed in Article 16 is further specified by a number of indicators, namely (i) affected users, (ii) length of the incident, (iii) "geographical spread" of the incident.

Because the Directive implements a so-called "light-touch approach", providers will have to comply to just one of the implementing acts across the different Member States, which is the one of the Country where they have their main establishment.

With regard to the dividing line between OESs and DSPs, ENISA explains that "significant incidents affecting OES networks and systems caused by incidents affecting DSPs' infrastructures are theoretically considered distinct incidents, therefore the reporting activities should also be distinct" [34]. Interestingly enough, public awareness is not a necessary step to be fulfilled to comply with the NIS Directive. Differently from what was mentioned above with regard to personal data breach notification requirements under the GDPR, in the realm of the NIS Directive, there is no obligation to keep the public informed about the incident unless public awareness "is necessary in order to prevent an incident or to deal with an ongoing incident, or where disclosure of the incident is otherwise in the public interest" [6].

In its guidance report [34], ENISA stresses the fact that the Directive requires incident reporting without undue delay. With respect to the adoption of information security standards (see for instance, ISO 27000) with a full set of protocols for incident handling and incident response, this approach might be on the one hand an additional layer of compliance and accountability, but on the other hand, following such protocols might be problematic with a view that incidents shall be promptly reported. Operators shall therefore try to strike a balance between reactiveness in informing competent authorities of any incident without undue delay and complying with international standards that include incident management within their realms.

---

[12]In chronological order.

Cooperation in the Field of Cybersecurity and Cyber Resilience.

In order to foster the exchange of information and enhance effective cooperation across Member States of the Union, the NIS Directive foresees the establishment of a so-called "Cooperation Group" [6], composed by the European Commission, ENISA and representatives of the Member States, tasked to cooperate with several Union bodies for the provision of advice and best practice on network security. According to Article 1.2(b), the scope of the Cooperation Group should be to foster information exchange and to develop trust and confidence amongst the Member States.

Amongst the numerous tasks, the Cooperation Group is in charge of a number of responsibilities which include providing active support to Member States in determining the organizations that fall under the definitions of Operators of Essential Services and Digital Service Providers (adopting a pan-European "consistent approach" [6]), hosting experts gathering for fostering the information sharing between actors.

With regard to the relationship between ENISA and the Cooperation Group, as stated in Recital 38, the two entities shall be deemed as "interdependent and complementary". This means that ENISA, in line with its establishing Regulation 526/2013, shall provide policy and analytical support as well as in the development of sector-specific guidelines. Additionally, implementing powers are given to the European Commission to set up the procedural arrangements for the Group.

National Cybersecurity Strategies.

Important domains within the NIS Directive do not solely pertain to the obligations that Member States shall impose on digital operators [25]. Rather, they also directly address Member States as such to start designing and developing a national strategy for cybersecurity (Article 7), which, according to Article 4(3), can be defined as a "framework providing strategic objectives and priorities on the security of network and information systems at national level". The aim of Article 7 is to indicate a twofold objective for each national cybersecurity strategy, i.e. (i) achieving and maintaining a high level of security of network and information systems and (ii) covering at least the sectors mentioned above (Operators of Essential Services and Digital Service Providers). The Directive sets up four main points that shall be addressed within each of the national cybersecurity strategy. Firstly, high-level priorities and objectives with regard to the security of information systems. Secondly, a governance structure that enables the achievements of such objectives. It needs to be noted that such a framework shall be rolling out quite granularly what roles and responsibilities will be given to national and governmental bodies with respect to such strategy. Thirdly, each national strategy shall set forth a series of measures aimed at addressing preparedness, response and recovery. Within this realm, specific mention is dedicated to the elicitation of rules and procedures for smooth cooperation between public bodies and private entities. Fourthly, strategies shall lay down ways and measures enabling education, awareness raising and training, as well as research and development programs in the area of cybersecurity and information and communication technology. Lastly, risk assessment plans and a list of actors involved in the implementation of the domestic cybersecurity strategy shall be explicitly mentioned in the national drafts.

Obligations on the Protection of Personal Data

Multiple overlaps can be identified not only between the NIS and the GDPR, but more broadly, between provisions, principles and obligations on privacy framework and cybersecurity. For instance, within the works of the Cooperation Group described above, the NIS makes explicit reference to the fact that, with regard to information and intelligence sharing between the Cooperation Group and amongst the various CSIRTs [35], the processing of personal data might be involved. Furthermore, as argued by many, national cybersecurity cooperation involving CSIRTs will help the governments to achieve a wider overview of the cyber threats scenarios, where such entities are considered as points of contact for an iterative exercise in the field of international cooperation. For this reason, interestingly enough, Recital 72 of the NIS Directive explains that any of such processing should be based on the standards imposed by a number of legal texts, namely:

- Regulation (EC) No 45/2001 [36] (now repealed by Regulation 2018/1725 [37]), to the extent that personal data are processed within the realm of European bodies and Institutions;
- Directive 95/46/EC [38], for any other personal data processing;
- Regulation (EC) No 1049/2001 [39], for the application of the NIS with regard to access to documents;

As we are able to see, no reference whatsoever to the upcoming GDPR is made in the list above, and the data protection standards mentioned therein still refer to the previous framework. With regard to potential overlaps within the subject matter, it needs to be noted that the NIS Directive explicitly differs between data protection authorities (DPAs) as opposed to competent authorities. According to Recital 63, in fact, such two, apparently separate entities, are asked to cooperate in the case a cyber-security incident has an impact on personal data (thus leading to a data breach) in the exchange of meaningful information and intelligence sharing. Such reasoning is confirmed by Article 8(6) which states that competent authorities (under Article 8(1), defined as information security national centers with single points of contact) shall establish lines of coordination along with DPAs and law enforcement agencies. Narrowly speaking, such cooperation is further detailed topic-wise. Recalling Recital 63 above in fact, Article 15(4) explicitly provides the legal basis for a cooperative collaboration between competent authorities and DPAs in the circumstance of incident reporting involving data protection aspects, most likely personal data breaches.

2.2   The Japanese Landscape

2.2.1   Privacy and Data Protection

The first recognition of the right to privacy in Japan comes with the jurisprudence in the Utage no Ato case [40], where, based on an interpretation of the Civil Code, the Tokyo District Court stated that "The right to privacy is recognized as the legal protection of the right so as not to be disclosed of private life" [41].

Following the period of political and social chaos deriving from the Second World War and the so-called national disasters of the Sixties, in 1975 the first claims for the right to privacy consolidated across the Japanese territories with the aim of limiting potential government surveillance and to give a response to the defeated national ID-scheme policies [42]. In the early 2000s, after the adaptation tendencies (with minor differences) of the Japanese legal landscape with parallel western countries [45], Japan realized that a legal approach to the right to privacy would have helped the nation's success in commerce and trade, with a substantially consensual approach of all the players involved: "The early signs from the business community, the bureaucracy, the political elite and the populace at large are overwhelmingly supportive of the conclusion that the Privacy Act will be effective" [42].

The Act on Personal Information came eventually into effect in 2005, after its official announcement in 2003. The Diet,[13] in that circumstance, received fierce opposition against the Act by the mass media community [40], and was constituted by two parts: basic principles (Chaps. 1 to 3), and general obligations (Chaps. 4 to 6).

The Japanese Reform on Privacy

The concept of privacy in Japan has rapidly evolved over the recent decades [46]. As briefly mentioned, Japanese data protection laws pertain to a relatively newer tradition in legal privacy [42]. Nonetheless Japan's first Act on Privacy (APPI—Act on the Protection of Personal Information), is one of "Asia's oldest data protection laws" [48].

The current privacy law got finalized in 2017 but the legislative process initiated some years back with a couple of relevant steps that the Japanese Government had to go through before formally starting the amendment of the existing laws. As Harada states, "One of the main objectives of the amended Act is to create a framework that recognizes and addresses the fact that transfers of personal data occur on a global scale" [49].

Particularly relevant to this end are in fact the Law Reform Plan adopted by the Cabinet IT Strategic Headquarter in 2012 and the Policy Outline of the Institutional Revision for Utilization of Personal data (2014). In the same year the Amendment Outline of the Bill was presented, and in March 2015 the formal process officially started with the Cabinet Decision on the Amendments on the Acts. The purpose of

---

[13]Parliamentary bodies of Japan.

the law reform was to foster innovation and to introduce new services as well as promoting people's safety by ensuring an appropriate protection of personal information. The first law is the Act on the Protection of Personal Information, which included amendments regarding both the processing and protection of personal data. It also established the independent regulatory authority by means of a significant restructuration of the existing information protection Commission.

Furthermore, in September 2015 the so called "My Number[14] Act" (ID Number Act) was set forth and included amendments on the promotion of the use of specific personal information (My Number) and on the numeration of the accounts, the use of numbers in the medical examination and health guidance, a link with vaccination records.

Japanese Privacy Act within the Japanese Legal System

It is important to note what is the structure of the Japanese legal system to better comprehend the role and the importance of the 2015 reform on privacy and data protection. To start with a top-down approach, the hierarchy of legal sources in Japan lists at its highest level the Constitution, under which a layer of so-called Fundamental Laws apply. Within such a cluster, the Act on the Protection of Personal Information can be considered as taking a similar role. The same act sees its Sects. 2.1, 2.2 and 2.3 standing a level below, where a number of legal principles are grouped, such as basic philosophy, responsibilities and measures of the State and Municipalities, or the establishment of the Basic Policy. Both acts on the Protection of Personal Information held by Administrative Organs and the one on the Protection of Personal Information retained by Incorporated Administrative Agencies occupy the lower levels, along with Sects. 4, 5, and 6 of the Act on Personal Information.[15]

Definition of Personal Information

According to Articles 2-1 of the Act, personal information is an information that is identifiable of the individuals by names, birthdate and the other descriptions including documents, drawings, electromagnetic records or voices, motions and other means.

Personal identifiers are instead described in Article 2-2; they are letters, numbers, marks and the other codes which fall in (i) characteristics of the part of body for the purpose of use of electronic machines, which is identifiable for the individual or, (ii) the individual user or purchaser designated, written or recorded in the service use or the sales.

Similarly to what we describe, according to the European privacy scholarship, as "special categories of personal data", Article 2-3 explains how personal information which include race, religious beliefs, social statuses, medical records, criminal

---

[14]From Wikipedia, Individual Number webpage: "An Individual Number, also known as My Number, is a 12-digit ID number issued to all citizens and residents of Japan (including foreign residents in Japan) used for taxation, social security and disaster response purposes. The numbers were first issued in late 2015".

[15]The authors would like to acknowledge the presentation on Japanese privacy laws by Prof. Hiroshi Miyashita (Chuo University) that took place under the auspices of the first EUNITY Workshop Meeting, held at the University of Tokyo in October 2017.

offences, events related to victims of criminal offences are indeed sensitive data, which require special care in order to avoid the injurious discrimination, biases and potentially subsequent disadvantages.

Conditions for Processing

It is noteworthy to highlight what the Japanese law says about anonymization. Specifically, Article 2-9 provides a definition for anonymous data as personal information which does not enable to identify an individual. It further explains two ways to achieve such a circumstance, (i) the deletion of all descriptions containing personal information and (ii) deletion of all personal identifiers, containing such information. This clause is supplemented by Article 2-10, regarding the anonymous processing information entities as any entity that uses the anonymous processing database (easily searchable via electronic machines).

For such cases, businesses will not have to seek for users' consent when they start processing data. Accordingly, such a measure was advocated by the Japanese Government to ease the shift of many digital business models towards big data analytics use and to facilitate their legitimization within the Japanese legal framework [48].

Data Broker Measures

According to article 25 of the Japanese Act, based on the Commission's rules on retention, (personal information handling) entities must keep records of the date and the third parties to which personal information are disclosed.

In the event that personal information handling entities receive the information from third parties, such entities shall check the basic details of the company (name, addresses and contacts of the representatives of the corporation), as well as the context under which such a transfer is made. Records shall be kept with regard to the date of data acquisition for a period of time set forth by the Commission.

In the event that personal information handling entities and / or their employees unlawfully acquire or steal personal information within a database for the purpose of making an illegal profit, the Act on the Protection of Personal Information provides for a punishment of up to one year imprisonment and up to 500.000 Yen under Article 83.

The Personal Information Protection Commission

Dating back in 2013, the My Number Act firstly established an independent oversight Commission with limited powers tailored for the monitoring and enforcement of such Act. The new Personal Information Act widens the scope of the pre-existing Commission as follows.

Chapters 4 and 5 of the Act on Personal Information provide for the basic rules and procedures of Japan's privacy regulator, called PIPC—Personal Information Protection Commission.

With regard to its establishment (Chap. 5), it should be noted that the PIPC falls under the jurisdiction of the Prime Minister's Cabinet Office, with a clear mission of ensuring the proper handling of personal data, taking into account the effective use of personal information. Article 52 enlists seven tasks of the Commission, namely:

1. Promoting the the Basic Policy
2. Supervising the use of My Number
3. Assessing the impact of My Number
4. Actively engaging in public relations and education
5. Conducting necessary studies and analysis
6. Becoming a player in the international cooperation field
7. Complying to the execution of other tasks provided for by law

The Commission performs its duties under the principle of independence of functions, particularly with regard to the President of the Commission and the Commissioners themselves. Having this in mind, it is useful to note that the governance of such body is composed by the President and eight Commissioners equally distributed in full-time and part-time posts. Such members are wholly appointed by the Japanese Prime Minister, with the approval of both Houses of the Diet (Japan's bicameral legislature). Their background can be of different disciplines, such as experts from Academia, consumer organizations, IT technologists, My Number Administrators, businesses and local organizations.

Their mandate lasts five years (renewable), and their status is guaranteed by immunity (unless they are involved in crimes related to insolvency, actions against the acts they supervise, crimes for which imprisonment is foreseen or they become mentally or physically disabled).

The Commission can establish specific expert committees to explore a particular issue or novelty, and has an obligation under Article 70 to report annually to the Diet. Section 3 (Chap. 4) of the Personal Information Act rolls out the main tasks of the Commission with regard to the supervision activities. According to such Section, the Commission has established powers of conducting on-site inspections and submitting reports. They can issue instructions and advices (Article 41), as well as recommendations and orders (Article 42).

Ultimately, the Commission has an explicit prohibition of non-interference with some of the basic civil rights such as the freedom of expression, academic freedom, freedom of religion and freedom of political association. On a number of occasions, it can delegate activities and powers to the Competent Minister which shall report to the Commission upon finalization of the task. Conversely, on a number of residual cases, the Competent Minister can request the Commission to take necessary measures in a given issue.

2.2.2 Japan and the European Union: Comparative Aspects on Privacy and Data Protection

Global Harmonization and Data Transfers

According to the Japanese Act, it is the task of the Government to take necessary measures to ensure the international harmonization with foreign states. Under Article 24, personal information cannot be transferred to a third party or a foreign country except in the cases of compliance with the Commission's standards or transfers to countries with an equivalent level of protection of interests and rights.

Similarly to the Japanese Act, the GDPR presents similar grounds for international cooperation in the area of personal data transfers with third countries. Article 45 states in fact that a mechanism for data transfers without the need of prior authorization or particular safeguards by the controllers is in place with those countries where an adequate level of data protection is ensured. Such an "adequacy decision" is the result of a relatively long assessment process (preceded by negotiations) undertaken by the European Commission, for which the following criteria serve as evaluation metrics: general (and specific) compliance with the rule of law and relevant civil rights, appropriate regulatory bodies in charge of supervision, enforcement and oversight and the international commitment the country has entered into by ways of any international public law instrument.

The European Union and Japan have signed an adequacy decision after the evaluation process by the European Commission [50]. Such a decision allows the free flow of personal data, and the mutual recognition that each other's territories abide to "essentially equivalent" standards for the protection of personal data. This process shall therefore comply to both the Japanese Personal Information Act (Article 24) and the GDPR (Article 45 and broadly, Chap. 5). The finalization of such evaluation leads to the issuing of an implementing act which, according to Article 45 of the GDPR, will be subject to a period review aimed at assessing significant changes and transformations happened in the meantime.

The smooth and legally authorized personal data flow between regions is crucial for fostering the digital trade market between such areas [51]. However, it needs to be kept in mind that the adequacy decision does only allow for free flow of personal data. Compliance with the foreign country rules where personal data needs to be transferred to (or from) will still have to be met through additional operational efforts and regardless of any such adequacy decision.

Comparative Overview of GDPR and Japanese Privacy Act

Even if a mutual recognition of adequate standards for data protection has entered into force within the biennium 2018–2019 [52], some differences between the two legal regimes still remain [53]. The comparison between the new reformed acts (GDPR and Japan's Personal Information Act) that is carried out below elicits a number of differences which give an example of the potentially significant operational efforts and compliance demands on businesses and organizations wanting to make benefit of the opportunity of both European and Japanese data markets [52].

Territorial Scope

With reference to the territorial scope, the GDPR details the extraterritoriality principle [5] (present in both jurisdictions) with an additional point. Such a principle
implies that the legal standards set up within the territory by the GDPR do also apply
outside such territories as long as foreign businesses and organizations offer goods
and services within the country or territory where the data protection laws apply.
Thus, both territories of the European Union and Japan extend the applicability of
their privacy laws to those established abroad but with the above-mentioned business interests within such regions. However, GDPR further extends such applicability
also to those firms which are not established within the EU but monitor and process
European data subject's behaviors.

Personal Information

Whilst the GDPR considers personal information as those relating to an identified
or an identifiable person [5], the Japanese Act differs in such definition as follows.
For such a concept in fact, the Japanese Personal Information Act provides a slightly
divergent notion, i.e. those information relating to a living individual which can
identify such specific subject by the description contained in the information.

Additionally, the Japanese Act adds up a definition of personal data, i.e. those
personal information that are processed within a database (with dedicated rules [52]
and procedures for such cluster of personal information).

Sensitive Personal Information

The list of sensitive personal information under the GDPR is comprised of those
revealing ethnic or racial origins, political opinions, religious or philosophical beliefs,
trade union membership and the processing of genetic or biometric data for identification purposes. The Japanese Act includes in its list information concerning religion,
social status or medical history, criminal records and the circumstance where a data
subject has suffered damages from a crime [52]. Such narrower interpretation of the
category of sensitive personal information by the Japanese laws has allegedly caused
some delays during the negotiations for the adequacy decision [51], for instance, the
European approach protects in this category trade union memberships and sexual
orientations, whilst the Japanese Act does not.

Information Rights

With respect to data subjects' rights, the Japanese Act distinguishes between the
duty of data controllers to provide for disclosure, rectification or interruption of
processing, added by the obligation of explanation of reasons (within the obligations
chapter). The GDPR enlists a number of more systematic rights, such as the rights
of information, access, rectification, erasure, objection and explanation. On top of
such basic rights, the GDPR improves user controls over his or her data by adding
a new right at Article 20 (currently absent in the Japanese provisions): the right to
data portability.[16]

---

[16]"The right to data portability allows individuals to obtain and reuse their personal data for their
own purposes across different services" [54].

Data Controllers and Processors

Unlike the European Union, which has a longstanding tradition in the use of the term "data controller" within its privacy laws, Japan does not seem to have a perfectly overlapping profile. Such concept is in fact replaced in the Personal Information Act by the notion of "business operator" [52], which is the responsible entity for the handling of the information. No further distinction with personal data processors are made in the law.

Penalties and Fines

The GDPR raises the bar for potential violation of its provisions up to 20M Euros or 4% of a business global annual turnover. However, no criminal liability arises from the regulation. Conversely, the Japanese Act has a lower breach sanctions cap (for instance, for database stealing, the fine is up to 500,000 Yen, roughly 4,200 Euros), while explicitly foreseeing up to one year imprisonment.

2.2.3 Cybersecurity

Since the early 2000s, Japan has been an active player in the cybersecurity field. A number of initiatives are to be recorded in this section, to set out the background that led to the legislation on cybersecurity which will further be elaborated below.

With regard to measures addressing **government agencies**, it needs to be noted that, in July 2000, a set of Guidelines on Information Security Policies was issued. Later on, between 2005 and 2014 a number of Common Standards for Information Security Measures for Government Agencies followed up the first Guidelines issued in 2000, with the 2014 edition published by the Policy Council in May 2014 [55].

With respect to the protection of **critical infrastructures**, the first initiative dates back to 2000 with the Special Action Plan on Cyber-terrorism Countermeasures for Critical Infrastructures, followed by two other Action Plans (respectively in 2005 and 2009) and the 2014 Basic Policy (3rd Edition), revised by the Strategic Headquarters in 2015 [55].

Regarding the **governance** model, in February 2000 the IT Security Office within the Cabinet Secretariat was established; after that in 2005 the National Information Security Center and the Information Security Policy Council were created [55].

Out of the merely legal and strategic initiatives, it is noteworthy to underline that Japan has undergone a number of serious cyber incidents. For instance, a survey conducted by KU Leuven within the auspices of the first EUNITY Workshop Meeting held in Tokyo in October 2017 by means of a questionnaire circulated amongst Japanese partners and attendants revealed that incidents like the Benesse data breach [56] (2014) had a significant impact on the cybersecurity community. However, the need for a structured and systematic response both at the strategic and at the legal level to cyber threats derives from a certain level of proactivity of the Japanese community towards the imminent Olympic Games, to be held in Tokyo in 2020 [57].

The Law

In November 2014, the Diet adopted the Cybersecurity Basic Act, amended in 2016 [58]. The Basic Act has a number of principles and notions in it; amongst others, it provides for a definition of "cybersecurity" in a legal terminology, the basic principles of cybersecurity policies, stakeholders' responsibility, structure and functioning of the Strategy and the Cybersecurity Headquarters composition. All such topic areas were therefore strengthened within the Japanese legal system by being formulated in one and only systematic legal text [59].

Moreover, the Cybersecurity Strategic Headquarters introduces mandatory reporting from government bodies and agencies, as well as the NISC[17] as a third, independent party which serves for auditing purposes and incident analysis. Furthermore, the Cybersecurity Strategy under the new Basic Act is no more an act adopted by the Information Security Policy Council (with no enforcement powers). Rather, it becomes a more accountable process, with the Act in itself being adopted with a Cabinet Decision and reported to the National Parliament, binding all governmental bodies and agencies and with clear enforcement powers [55].

A Legal Definition for Cybersecurity

First and foremost, the Basic Act defines cybersecurity within the realm of the Japanese Law. Accordingly, it is intended as "consideration, maintenance and management of needed measures to prevent the leakage, destruction or damaging of information reported or transmitted or received by electronic way, magnetic way or other ways that human cannot recognize, or to manage safety control of that information, or to ensure safety and reliability of information systems or information and communication networks" [47]. Interestingly, the European Union's Network and Information Security Directive (NIS) does not define what cybersecurity is within the context of the Directive itself.

NIS versus Basic Act: the Citizen as an Active Stakeholder

It is important to highlight in this section a difference between the Basic Act and the European NIS Directive. Whilst the latter merely represents an attempt to coordinate efforts towards the security of European information systems by defining a set of rules that need to be implemented and that address a limited amount of players, Article 9 of the Basic Act on Cybersecurity [47] does actually mention the responsibilities of the citizens within the cybersecurity domain: "In accordance with the Basic Principles, the people are to make an effort to deepen their awareness and understanding of the critical value of Cybersecurity and pay necessary attention to ensuring Cybersecurity".[18]

It is not fully clear whether such a provision creates a positive obligation of a duty of care for the citizens or if it is merely a programmatic norm which remarks the concept summed by the Latin clause ignorantia juris non excusat (tr., ignorance of law excuses no one). However, what is striking in the Basic Act is the presence

---

[17]National center of Incident readiness and Strategy for Cybersecurity, established in January 2015.
[18]Tr. "Japanese law translation".

of the citizen as an active stakeholder, a player from whom some level of awareness is expected. To this end, one of the so-called Basic Principles, Article 3(2), invites all citizens to put in place voluntary actions aimed at preventing damages caused by cyber threats. This is not only relevant from a legal perspective. Rather, it reveals the level of maturity of a legislation and of a society which tends to include the citizen within the scope of such a legal text on the above-mentioned specific domain, whilst its European alter ego (NIS), limits itself in addressing a restricted number of players. Notwithstanding the fact that there is a substantial difference in the two legal texts (one, the Basic Act, is directly applicable, while the NIS Directive needs an active implementation act by Member States), it should be nevertheless noted how the Japanese approach seems to mirror the general claim by the expert community that cybersecurity is a topic that has to be grasped and addressed also at the societal level.

Information Sharing and Incident Reporting

"Ahead of the 2020 Olympic Games in Tokyo, the government sees cooperation as necessary to better prepare for growing cybersecurity threats, recover from such damage and probe the causes". This is how a government senior official was addressing the Japan Times [60] on cybersecurity cooperation. As already mentioned earlier, the Japanese Government has proved to be willing to approach large events with potential cyber implications proactively.

In such circumstance, the point was that the Japanese Government was set to start filing a cybersecurity cooperation request according to the Basic Act, addressing 48 entities that were identified on the basis of the Basic Act itself, and for which it is foreseen a coordinated cooperative action to counter potential cyber threats before, during and after the Olympic Games.

With regard to information sharing, the Basic Act does not contain a coherent provision. Rather, it includes three different layers: Article 13, with respect to the measures to be provided to cybersecurity players such as administrative organs by the Government, which includes responses to cybersecurity threats and cooperation, communication and coordination with relevant domestic and foreign parties and sharing of information about cybersecurity across all sectors of the governmental machinery.

As a second layer, Article 14 of the Basic Act includes a provision requiring the Government to activate and set up the necessary measures in order to enable the promotion of information sharing. Lastly, Article 23 enlists a series of explicit areas where the Government is expected to carry out Japan's role in the field of cybersecurity. These are measures like active participation in law-making of international norms, confidence and trust building towards the exchange of information with foreign countries, international technical cooperation on capacity building and cybercrime takedowns, awareness raising across the international community.

Over the recent years, Japan has signed a number of agreements on information sharing with regard to cybersecurity threat intelligence with a number of partners, which are not only located in the Asia-Pacific area, rather all over the world, as a symptom that this domain needs a global cooperative platform in which, for example, the relationships between the EU and Japan can be significantly improved. Examples of such partnerships are (non-exhaustively) listed hereby:

1. Japan-UK Joint Declaration on Security Cooperation, Tokyo, (2017)[19]
2. Japan-India Cyber Dialogues, which lasts since 2012 and is at its Second Edition (2017)[20]
3. The Japan-ASEAN cybersecurity dialogues, which are rolled out in a number of different policy initiatives since 2009 [61].
4. Japan's first access to the US's DHS's Automated Indicator Sharing (AIS) [62].

2.3 Conclusions

The present chapter highlights the main aspects of the two regional entities, the European Union and Japan, as well as their progress in the legal and regulatory area within the privacy and cybersecurity domain.

Both regions have undergone a period of substantial reforms and modernization of the existing legal frameworks with regard to privacy. In Europe, the GDPR entering into force in May 2018 brings up a significant update of the legislation with a strict compliance bar foreseen for its enforcement. In Japan, the Act on Personal Information was recently amended. However, the so-called adequacy decision for the free mutual transfer of personal data has been adopted between the EU and Japan, which allows a mutual recognition of an equivalent level of data protection. It should be nevertheless noted that the two frameworks are not perfectly matching. For instance, the concepts of sensitive personal data, as well as some practical implications such as the diverging approach on the imposition of fines and sanctions might become a critical point for both Japanese and European businesses and organizations wanting to enter each other's digital markets.

In contrast, the cybersecurity domain is somewhat dissimilar from the one described above. At a first glance, differences might be spotted in the laws of the two regions, mostly pertaining to their scope and the stakeholders such laws apply to, thus diverging from one another in substantial elements. However, some lines of similarities can yet be pointed out. This can be observed in the room left by both policy and legal frameworks allowing EU, Member States and Japanese Government to engage in international cooperation, particularly in the fields of international norms settings and common strategy building, promotion and awareness raising of a cybersecurity culture within the respective territories and beyond.

---

[19]For more information, consult: [43].

[20]For more information, consult: [44].

2.3.1 Summary of Challenges and Gaps

This conclusive section provides for a number of points that have been identified as potential blockers or issues to be addressed in order to further strengthen the cooperation between the EU and Japan in the domains of privacy and cybersecurity.

(a) The **scope** of the two privacy legislations differ quite significantly in their application: whilst Japanese laws include the offering of goods and services only, the EU framework extends its scope to non-European organizations offering services in Europe and to those targeting market behaviors of data subjects. For this reason, the competitive imbalance may stand in favor of the European side, seemingly creating a burden on Japanese organizations wanting to monitor market behaviors of their potential customers.

(b) The **concept of personal information** is not aligned between the two regions. Whilst the EU takes a more protective and data-subject-driven perspective, (inter alia, by including within such a definition IP addresses), the Japanese standpoint is more restrictive in the delineation of the contours of such definition, thus creating a potential gap. The result could be that Japanese businesses processing information of EU data subjects will have to abide to data protection laws for processes that are normally excluded by such compliance in their country.

(c) **Information rights** result to be more articulated in the EU framework, entailing much more options for data subjects to control their personal data. For this reason, Japanese controllers would have to put in place processes to permit a number of actions (see portability for instance), which in contrast are not mandatory in the Japanese framework.

(d) The **sanctionatory regime** appears higher in the European legislation than the one in Japan. Therefore, assessment of the advantages in entering the other region's market would definitely result more advantageous for Europeans offering services in Japan than vice versa, as far as the mere evaluation of the risk of being fined is concerned. However, the Japanese law entails some aspects of criminal liability (with prison penalties in some residual cases of privacy violations), which are not present in European law.

(e) With regard to **cybersecurity**, the Cyber Security Act is expected to provide substantial improvement to the current European legal framework on cybersecurity, which was previusly solely constituted by the NIS Directive. However, one of the major differences observed in the comparison of the two frameworks is the fact that Japan actively addresses the conduct of the citizens with respect to cybersecurity. Such *duty of care*, lifts the burden of protecting information systems that in Europe merely stands on producers, manufacturers and other stakeholders in the supply chain.

(f) With regard to **information sharing**, a positive note is coming from the adequacy decision being signed between the EU and Japan. This will allow a mutual recognition of an equivalent level of data protection by the EU and Japan. Once adopted, this will cover *"personal data exchanged for commercial purposes, but also personal data exchanged for law enforcement purposes between EU and*

Japanese authorities, ensuring that in all such exchanges a high level of data protection is applied" [63]. The press release from the European Commission (17 July 2018) continues:

To live up to European standards, Japan has committed to implementing the following additional safeguards to protect EU citizens' personal data, before the Commission formally adopts its adequacy decision:

a. *A set of rules providing individuals in the EU whose personal data are transferred to Japan, with additional safeguards that will bridge several differences between the two data protection systems. These additional safeguards will strengthen, for example, the protection of sensitive data, the conditions under which EU data can be further transferred from Japan to another third country, the exercise of individual rights to access and rectification. These rules will be binding on Japanese companies importing data from the EU and enforceable by the Japanese independent data protection authority (PPC[21]) and courts.*

b. *A complaint-handling mechanism to investigate and resolve complaints from Europeans regarding access to their data by Japanese public authorities. This new mechanism will be administered and supervised by the Japanese independent data protection authority* [63].

2.3.2 Policy Blockers

On a broader scale and perspective, the advent of the GDPR and the continuous reforms in terms of privacy and cybersecurity seem not to have influenced the processing of personal data solely within the territories of the European Union. Rather, its scope and impact force us to reflect on the potential effects that such a Regulation may have on the broad, global business at large.

Whilst many commentators brought forward the undoubtable benefits of the GDPR on the strengthening of the protection of personal data as a high standard for information security practices, it is worth mentioning that such requirements might become obstacles for services and businesses that operate on a much wider scale than the mere regional ones (EU or Japan, for instance).

Under such auspices, the following examples describe those businesses and practices that might be impacted by the policies implemented by the GDPR and its Japanese counter-part, APPI. The selection of such examples took into account two of the major cases on internet governance and related services with a significant jurisprudential importance for privacy and data protection in a GDPR context.

(a) **Clarification on the adequacy decision and review mechanisms**. From a data protection perspective, an adequacy decision has recently been signed between the two regions. This will allow for the creation of the biggest space for digital

---

[21] Japanese privacy regulatory body.

exchange worldwide. It goes without saying that rules and procedures shall not only be adequate and in place, but also operationalized and contextualized properly. The EDPB, in its opinion on the Adequacy Decision, remarks a number of recommendations to be implemented prior and after the decision is signed. In the future, it will be important that mechanisms for the review and the monitoring of such a wide digital marketspace enable the efficiency of both digital innovation and enforcement of digital rights.

(b) **Processing of IP addresses**. From a privacy perspective, the European Union data protection framework considers IP addresses as personal information, hence falling under the legal regime disciplined by the EU privacy reform package. In a EU-Japanese collaboration perspective, this could become a potential blocker for future projects which include the collection and the processing of IP addresses of European citizens by Japanese businesses or research entities. Cybersecurity, telecommunication and market research are only three of the sectors which could be impacted by the extensive scope of the GDPR. The misalignment between Japan and the European union on the definition of personal information therefore substantiates with a great extent and magnitude in the case of IP addresses. Such a difference in legal interpretation may become a great obstacle in terms of future cooperation in the research domain, as well as in the provision of digital services and the achievement of future partnerships between the two regions.

(c) **National GDPR implementations**. Whilst the GDPR is a strong legal instrument which imparts automatic implementation to the EU Member States, a number of provisions in its text still needs national specification. See for instance, the age for consent to the processing of personal data by minors: each Member State has to establish an age on its own, and lay it down in the national implementation act of the GDPR. This, alongside a number of other exceptions derogated to Member States' decision (see also the limits and scope of the research exemption), may create legal fragmentation and uncertainty, which may constitute, particularly from the Japanese side, a blocker in the engagement of business and policy initiatives between the two regions and between Japan and national Member States.

(d) **WHOIS**. With the advent of the GDPR, the public availability of the notorious WHOIS databases got seriously questioned by both civil society organizations and privacy regulators. Even though this discussion had been going on for a very long time (starting a decade earlier than the GDPR), it is with the privacy reform that happened at the EU level a couple of years ago (2016) that ICANN—Internet Corporation for Assigned Names and Numbers—the entity coordinating the WHOIS databases policies, started to impose stricter rules on such a service. It is worth mentioning that ICANN does not directly run and manage such lists, which in turn are the result of a trade-off between registries and registrars (they are in fact commercially exploited). Under the new GDPR regime, ICANN decided to amend its policies demanding to WHOIS database owners to implement a version of the list without a significant number of identifiers, which were instead present before. Furthermore, ICANN is developing a privileged access option for those actors with a legitimate interest to access such data. All these new mechanisms are set to have a global application to all ICANN-related activities. European provisions will therefore change the way the entire internet governance will work, thus impacting also all other regions of the globe, potentially including Japan. Sectors using WHOIS databases for their daily activities include public actors (law enforcement agencies), but also a very wide part of the private security industry (cybersecurity firms and providers, intellectual property lawyers to name a few).

(e) **Cybersecurity information sharing**. Strictly speaking, the cybersecurity regulatory regimes of the two regions are not mutually recognized. Whilst the privacy framework will provide for an adequacy decision for the exchange of personal data between the EU and Japan, the same cannot be yet said for the exchange of threat intelligence between CERTs. This might create legal uncertainty, which may well hamper the need of promptness in information sharing, particularly in crisis management.

(f) **Concerns around the respect for the human to the right to privacy in Japan in criminal intelligence, investigation and prosecution of terrorist case**. On a broader scale, issues with regard to the adequate standard of protection of privacy as a human right has been raised and may constitute a potential blocker for future policy collaboration between the two countries. The so-called "Act on Punishment of the Preparation of Acts of Terrorism and Other Organized Crimes" in fact has caught the attention of the United Nations' Special Rapporteur on the Right to Privacy, Prof. Joseph Cannataci. Accordingly, such a bill, due to its rather broad and unspecified scope, may lead to "undue restrictions to the rights to privacy and to freedom of expression". The bill, which criminalizes a great number of non-criminal activities as 'preparatory acts' under the terrorism label, leads to the intuition that a significant increase of digital surveillance may occur on Japanese individuals and their online activities. This may include their e-commerce records and transactions, thus involving the surveillance of businesses of individuals offering goods and services, too.

References

1. Cuijpers, C., Koops, B.-J.: How fragmentation in European law undermines consumer protection: the case of location-based services (2010). Eur. Law Rev, **33**, 880-897 (2008). https://ssrn.com/abstract=1645524. Cited 1 Jul 2019
2. Ehrenfeld, J.: WannaCry, cybersecurity and health information technology: a time to act. J. Med. Syst. **41**(7), 1 (2017)
3. Kuner, C.: Privacy, security and transparency: challenges for data protection law in a New Europe. Eu. Bus. Law Rev. **16**(1), 1–8 (2005)
4. Tankard, C.: What the GDPR means for businesses. Netw. Secur. **2016**(6), 5–8 (2016)
5. Regulation (EU) 2016/679 of the European Parliament and of the Council of 27 April 2016 on the protection of natural persons with regard to the processing of personal data and on the free movement of such data, and repealing Directive 95/46/EC (General Data Protection Regulation). https://eur-lex.europa.eu/eli/reg/2016/679/oj. Cited 1 Dec 2019

6. Directive (EU) 2016/1148 of the European Parliament and of the Council of 6 July 2016 concerning measures for a high common level of security of network and information systems across the Union. https://eur-lex.europa.eu/eli/dir/2016/1148/oj. Cited 1 Dec 2019

7. Directive 95/46/EC of the European Parliament and of the Council of 24 October 1995 on the protection of individuals with regard to the processing of personal data and on the free movement of such data

8. Directive 2002/58/EC of the European Parliament and of the Council of 12 July 2002 concerning the processing of personal data and the protection of privacy in the electronic communications sector (Directive on privacy and electronic communications). https://eur-lex.europa.eu/legal-content/EN/ALL/?uri=CELEX%3A32002L0058. Cited 1 Dec 2019

9. De Hert, P., Czerniawski, M.: Expanding the European data protection scope beyond territory: article 3 of the General Data Protection Regulation in its wider context. In. Data Priv. Law **6**(3), 230–243 (2016)

10. Technology neutrality in Internet, telecoms and data protection regulation. Hogan Lovells Global Media and Communications Quarterly 2014. http://www.hoganlovells.com/files/Uploads/Documents/8%20Technology%20neutrality%20in%20Internet.pdf. Cited 21 Feb 2018

11. Ausloos, J., Dewitte, P.: Shattering one-way mirrors. Data subject access rights in practice. Int. Data Priv. Law **8**(1). https://ssrn.com/abstract=3106632. Cited 20 Jan 2018

12. van Bastelaere, D.: GDPR-How EU citizens get back control of their personal data. Tri-Finance Blog (2017). https://blog.trifinance.com/gdpr-eu-citizens-get-back-control-personal-data/. Cited 15 Jan 2018

13. Charter of Fundamental Rights of the European Union. Article 8. https://www.europarl.europa.eu/charter/pdf/text_en.pdf. Cited 1 Dec 2019

14. ENISA: Recommendations on European Data Protection Certification (2017). https://www.enisa.europa.eu/publications/recommendations-on-european-data-protection-certification/at_download/fullReport. Cited 1 Dec 2019

15. ENISA: Looking into the crystal ball: a report on emerging technologies and security challenges (2018). https://www.enisa.europa.eu/publications/looking-into-the-crystal-ball/at_download/fullReport. Cited 1 Dec 2019

16. Regulation (EC) No 765/2008 of the European Parliament and of the Council of 9 July 2008 setting out the requirements for accreditation and market surveillance relating to the marketing of products and repealing Regulation (EEC) No 339/93. https://eur-lex.europa.eu/LexUriServ/LexUriServ.do?uri=OJ:L:2008:218:0030:0047:EN:PDF. Cited 1 Dec 2019

17. Article 29 Data Protection Working Party: Guidelines on Data Protection Impact Assessment (DPIA) and determining whether processing is "likely to result in a high risk" for the purposes of Regulation 2016/679 http://ec.europa.eu/newsroom/document.cfm?doc_id=47711. Cited 1 Dec 2019

18. Article 29 data protection working party: guidelines on personal data breach notification under Regulation 2016/679. https://ec.europa.eu/newsroom/article29/document.cfm?action=display&doc_id=49827. Cited 1 Dec 2019

19. Article 29 data protection working party: Opinion 03/2014 on personal data breach notification https://ec.europa.eu/justice/article-29/documentation/opinion-recommendation/files/2014/wp213_en.pdf. Cited 1 Dec 2019

20. Communication from the Commission to the European Parliament, the Council, the European Economic and Social Committee and the Committee of the Regions on the Mid-Term Review on the implementation of the Digital Single Market Strategy - A Connected Digital Single Market for All. COM/2017/0228 final. https://eur-lex.europa.eu/legal-content/EN/TXT/?uri=CELEX%3A52017DC0228. Cited 1 Dec 2019

21. Leszczyna, R.: Standards on cyber security assessment of smart grid. Int. J. Crit. Infrastruct. Prot. (2018)

22. Regulation (EC) No 460/2004 of the European Parliament and of the Council of 10 March 2004 establishing the European Network and Information Security Agency. https://eur-lex.europa.eu/legal-content/EN/TXT/?uri=CELEX%3A32004R0460. Cited 1 Dec 2019

23. European Commission proposal on a Regulation of the European Parliament and of the Council on the future of ENISA. https://www.enisa.europa.eu/news/enisa-news/european-commission-proposal-on-a-regulation-on-the-future-of-enisa. Cited 1 Dec 2019

24. Proposal for a Regulation of the European Parliament and of the Council on ENISA, the "EU Cybersecurity Agency", and repealing Regulation (EU) 526/2013, and on Information and Communication Technology cybersecurity certification ("Cybersecurity Act"). COM/2017/0477 final - 2017/0225 (COD). https://eur-lex.europa.eu/legal-content/EN/TXT/?qid=1505290611859&uri=COM:2017:477:FIN. Cited 1 Dec 2019

25. Europe emerges as global leader in cybersecurity law enforcement to protect critical infrastructure. PRNewswire. (2016). https://www.prnewswire.com/news-releases/europe-emerges-as-global-leader-in-cybersecurity-law-enforcement-to-protect-critical-infrastructure-300285914.html. Cited 1 Dec 2019

26. Annual report from the High Representative of the European Union for Foreign Affairs and Security Policy to the European Parliament : Main aspects and basis choices of the CFSP. (2014). https://library.euneighbours.eu/sites/default/files/attachments/st14924_en.pdf. Cited 1 Dec 2019

27. Moret, E., Pawlak, P.: The EU Cyber Diplomacy Toolbox: towards a cyber-sanctions regime?. EUISS, Brief Issue. https://www.iss.europa.eu/sites/default/files/EUISSFiles/Brief%2024%20Cyber%20sanctions.pdf. Cited 1 Dec 2019

28. Bull, K.: New law to improve critical infrastructure cybersecurity. Security, **53**(9), 20-20,22 (2016)

29. Jenkins, J., Forsyth, C., Amole, A.: Should cyber security standards be imposed by regulation or left to discretion? Legal Week **16**(3), 10 (2014)

30. Weiss, J., Webb, B.: Standards coordination for cyber security. Intech **52**(12), 76 (2005)

31. Regulation (EU) No 1025/2012 of the European Parliament and of the Council of 25 October 2012 on European standardisation, amending Council Directives 89/686/EEC and 93/15/EEC and Directives 94/9/EC, 94/25/EC, 95/16/EC, 97/23/EC, 98/34/EC, 2004/22/EC, 2007/23/EC, 2009/23/EC and 2009/105/EC of the European Parliament and of the Council and repealing Council Decision 87/95/EEC and Decision No 1673/2006/EC of the European Parliament and of the Council. L 316/12. Article 2. https://eur-lex.europa.eu/eli/reg/2012/1025/oj. Cited 1 Dec 2019

32. Bøgh, S.A. (Ed).: A world built on standards–A textbook for higher education. ISBN: 978-87-7310-963-2. Danish Standards Foundation (2015)

33. Stuurman, K., Irene, K.: IoT Standardization-The Approach in the Field of Data Protection as a Model for Ensuring Compliance of IoT Applications?. In: 2016 IEEE 4th International Conference on Future Internet of Things and Cloud Workshops (FiCloudW), pp. 336-341 (2016)

34. ENISA: Incident notification for DSPs in the context of the NIS Directive (2017). https://www.enisa.europa.eu/publications/incident-notification-for-dsps-in-the-context-of-the-nis-directive/at_download/fullReport

35. Schweighofer, E., Heussler, V.Kieseberg, P.: Privacy by design data exchange between CSIRTs. In: Schweighofer, E. Leitold, H. Mitrakas, A. Rannenberg, K. (eds) Privacy Technologies and Policy. APF, : Lecture Notes in Computer Science, vol. 10518, p. 2017. Springer, Cham (2017)

36. Regulation (EC) No 45/2001 of the European Parliament and of the Council of 18 December 2000 on the protection of individuals with regard to the processing of personal data by the institutions and bodies of the Community and on the free movement of such data. OJ L 8. https://eur-lex.europa.eu/eli/reg/2001/45/oj. Cited 1 Dec 2019

37. Regulation (EU) 2018/1725 of the European Parliament and of the Council of 23 October 2018 on the protection of natural persons with regard to the processing of personal data by the Union institutions, bodies, offices and agencies and on the free movement of such data, and repealing Regulation (EC) No 45/2001 and Decision No 1247/2002/EC. OJ L 295, 21.11.2018. https://eur-lex.europa.eu/legal-content/EN/ALL/?uri=CELEX%3A32018R1725. Cited 1 Dec 2019

38. Directive 95/46/EC of the European Parliament and of the Council of 24 October 1995 on the protection of individuals with regard to the processing of personal data and on the free

movement of such data. No L 281/31. https://eur-lex.europa.eu/eli/dir/1995/46/oj. Cited 1 Dec 2019

39. Regulation (EC) No 1049/2001 of the European Parliament and of the Council of 30 May 2001 regarding public access to European Parliament, Council and Commission documents. OJ L 145. https://eur-lex.europa.eu/eli/reg/2001/1049/oj. Cited 1 Dec 2019

40. Miyashita, H.: The evolving concept of data privacy in Japanese law. Int. Data Priv. Law **1**(4), 229–238 (2011). https://doi.org/10.1093/idpl/ipr019

41. Judgment of Tokyo District Court, 28 September 1964, Hanrei-jiho, vol. 385, p. 12

42. Lawson, C.: Japan's New Privacy Act in Context, 29 U.N.S.W.L.J. 88 (2006)

43. Japan-UK Joint Declaration on Security Cooperation. (2017). https://assets.publishing.service. gov.uk/government/uploads/system/uploads/attachment_data/file/641155/Japan-UK_Joint_ Declaration_on_Security_Cooperation.pdf. Cited 1 Dec 2019

44. Joint Press Release Second Japan-India Cyber Dialogue. New Delhi. (2017). https://www. mofa.go.jp/press/release/press4e_001698.html. Cited 1 Dec 2019

45. Oda, H.: Japanese Law (1992)

46. Miyashita, H.: changing privacy and data protection in Japan. 10 The Sedona Conf. J. 277, 278 (2009)

47. Basic Act on Cybersecurity, Act No. 104 of November 12, 2014 (amended 2016)

48. Pearson, Harriet. Brill, Julie. Parsons, Mark and Imai, Hiroto: Changes in Japan Privacy Law to Take Effect in Mid-2017; Key Regulator Provides Compliance Insights. Hogan Lovells. (2017). https://www.lexology.com/library/detail.aspx?g=efa0a2b0-b73e-456c-b4fa-26a268e9e751. Cited 1 Dec 2019

49. Harada, M.: Japan: Personal data protection. Int. Financ. Law Rev. (2017)

50. Communication from the Commission to the European Parliament and the Council on Exchanging and Protecting Personal Data in a Globalised World. COM/2017/07 final/2. https://eur-lex. europa.eu/legal-content/EN/TXT/?uri=CELEX%3A52017DC0007R%2801%29. Cited 1 Dec 2019

51. Fioretti, J.: EU sees data transfer deal with Japan early next year. Reuters. (2017). http://news. trust.org/item/20171215124551-cd0x4. Cited 1 Dec 2019

52. Takase, K.: GDPR matchup: Japan's Act on the protection of personal information. IAPP (2017). https://iapp.org/news/a/gdpr-matchup-japans-act-on-the-protection-of-personal-information/. Cited 1 Dec 2019

53. Scott, M., Cerulus, L.: Europe's new data protection rules export privacy standards worldwide. Politico.eu. (2018). https://www.politico.eu/article/europe-data-protection-privacy-standards-gdpr-general-protection-data-regulation/. Cited 1 Dec 2019

54. The Information Commissioner's Office (ICO): Right to data portability. https://ico.org.uk/ for-organisations/guide-to-data-protection/guide-to-the-general-data-protection-regulation-gdpr/individual-rights/right-to-data-portability/. Cited 1 Dec 2019

55. Yamauchi, T: Cybersecurity strategy in Japan. (2017). https://project.inria.fr/ FranceJapanICST/files/2017/05/TYamauchi_presentation_2017.pdf. Cited 1 Dec 2018

56. Nikkei Asian review: Customer data leak deals blow to Benesse. (2014). https://asia.nikkei. com/Business/Customer-data-leak-deals-blow-to-Benesse. Cited 1 Dec 2019

57. Kelly, T., Kubo, N.: Japan holds first broad cybersecurity drill, frets over Olympics risks. Reuters. (2014). https://www.reuters.com/article/us-japan-cybercrime/japan-holds-first-broad-cybersecurity-drill-frets-over-olympics-risks-idUSBREA2G1O920140317. Cited 1 Dec 2019

58. Sayuri, U.: Japan: Cybersecurity basic act adopted, global legal monitor. (2014). https:// www.loc.gov/law/foreign-news/article/japan-basic-act-on-cybersecurity-amended/. Cited 1 Dec 2019

59. Kim, K., Park, S.Lim, J.: Changes of cybersecurity legal system in East Asia: Focusing on comparison between Korea and Japan. In: Kim, H. Choi, D. (eds) Information Security Applications. WISA, : Lecture Notes in Computer Science, vol. 9503, p. 2016. Springer, Cham (2015)

60. 48 infrastructure entities to get cybersecurity cooperation requests. The Japan Times (2015). https://www.japantimes.co.jp/news/2015/03/01/national/crime-legal/48-infrastructure-entities-to-get-cybersecurity-cooperation-requests/. Cited 1 Dec 2019

61. Matsubara, M.: Japan's Cybersecurity Capacity-Building Support for ASEAN–Shifting From What to Do to How to Do It. PaloAlto Networks. (2017). https://blog.paloaltonetworks.com/2017/07/cso-japans-cybersecurity-capacity-building-support-asean-shifting/. Cited 1 Dec 2019

62. Chalfant M.: US, Japan deepen cyber information sharing. (2017). https://thehill.com/policy/cybersecurity/331979-us-japan-deepen-cyber-information-sharing. Cited 1 Dec 2019

63. European Commission: "The European Union and Japan agreed to create the world's largest area of safe data flows". EU News 176/2018. (2018). https://eeas.europa.eu/delegations/japan/48487/european-union-and-japan-agreed-create-worlds-largest-area-safe-data-flows_en. Cited 1 Dec 2019

Chapter 3
Research and Innovation Aspects

This chapter contains the description of the most important research and innovation aspects regarding cybersecurity and privacy. Sections 3.1 and 3.2 have mainly descriptive character, and regard the area of research funding and the area of substantive research directions, respectively. Sections 3.3 and 3.4, in contrast, provide some conclusions regarding research in cybersecurity and privacy and also contain some suggestions about future common interests between the European Union and Japan.

3.1 Mechanisms to Finance Cybersecurity Research

This section identifies and describes the mechanisms to finance cybersecurity research in both regions.

3.1.1 In the European Union

This subsection identifies and describes the mechanisms to finance cybersecurity research in the European Union. We start by generally describing the funding mechanisms in the European Union, followed by a detailed description of funding mechanisms on the country level on the basis of Polish example.

3.1.1.1 European Union, International

Framework Programmes (now Horizon 2020)—open competitions

- Applicant: legal entities established in Member States, overseas territories linked to Member States, associated countries, or listed countries

© The Author(s), under exclusive license to Springer Nature Switzerland AG 2021
A. Felkner et al., *Cybersecurity Research Analysis Report for Europe and Japan*,
Studies in Big Data 75, https://doi.org/10.1007/978-3-030-62312-8_3

- Project manager: not specified
- Duration: may be specified in topic
- Level of financing: 70–100%
- Budget:

  - programme: specified in Work Programme
  - call: specified per topic (per strand if applicable)
  - project: specified approximated contribution per project but this is not required

- TRL: specified per topic (per strand if applicable)

Special programmes such as Connecting Europe Facility (CEF)—Cybersecurity *calls*

- Applicant: one or more Member State, international organisations, joint undertakings, public or private undertakings
- Project manager: not specified
- Duration: 24 months
- Level of financing: 75%
- Budget:

  - programme: not specified
  - call: up to EUR 12 million
  - project: specified approximated contribution per project

- TRL: not specified

ENISA-European Union Agency for Network and Information Security
Announces tender procedure for multiple framework contracts with re-opening competition. The budget for each framework contract period is specified in the tender. Successful tenderers sign framework contracts and are invited to submit their offers in specific tenders.

EUREKA

Has been created as an intergovernmental initiative in 1985 and aims at enhancing European competitiveness through its support to pan-European projects to develop innovative products, processes and services. EUREKA promotes and supports market-oriented international R&D&I project generation and facilitates access to finance for companies involved in its projects. The main aim of EUREKA is: *"raising the productivity and competitiveness of European businesses through technology. Boosting national economies on the international market, and strengthening the basis for sustainable prosperity and employment."* [1]

- Applicant: research centres, corporations, administrations, SMEs, universities
- Regions of applicants: European Union, Mediterranean countries, Switzerland, Balkans and other countries.

---

[1] http://www.eurekanetwork.org.

EUREKA is a framework programme and is organized around 4 different instruments :

- **EUREKA Network Projects** is an instrument focused on transnational, market-driven innovative research and development projects. Projects are supported by the public administrations and public funding agencies that represent EUREKA. Projects aim to develop marketable products, services or processes.
- The **Eurostars** Programme is a joint programme between EUREKA and the European Commission and the first European funding and support programme to be specifically dedicated to research-performing SMEs. Eurostars stimulates them to lead international collaborative research and innovation projects by providing support and funding. Eurostars projects are collaborative, meaning they must involve at least two participants (legal entities) from two different Eurostars participating countries.
- EUREKA **Clusters** are long-term and strategically significant initiatives that develop technologies. A typical Cluster project is performed by 2 to 14 participants from 2 to 4 countries.
- EUREKA **Umbrellas** are an association of at least five national thematic networks in a specific field of technology or service. Umbrellas are comprised of experts from science and industry as well as representatives of national funding institutions. They are set up to support and advise project consortia on developing project ideas through international partnerships in their given field of technology.

Other Forms of Financing Mechanism:

- Marie Sklodowska-Curie actions/grants and support for research infrastructures
- Projects of the European Strategy Forum for Research Infrastructures (ESFRI)
- European Structural and Investment Funds (ESIF)
- PRIMA initiative (Euro-Mediterranean cooperation)
- Euratom (European Atomic Energy Community) programs which support projects connected to nuclear energy (which could include cybersecurity aspects).

3.1.1.2 Financing Mechanism in the Country Level on the Example of Poland

In Poland cybersecurity research may be financed by various sources, both national and international. International / European Union programs are described in the previous Sect. 3.1.1.1. The national funding programs are described below.
National programmes

The National Centre for Research and Development (NCBR)—implementing agency of Minister of Science and Higher Education, [2] for example:

---

[2]http://www.ncbr.gov.pl/en/.

- **National Defence and Security Calls** (both open and closed) usually single-stage calls.[3] In each edition catalogue of eligible costs slightly differed from the previous ones.

  - call 9/2018 on performing and financing projects in the area of scientific research and development for defence and security of the state

    - Applicant: research consortium/ research-industrial centre/ entrepreneurs
    - Project manager: not specified
    - Duration: not specified
    - Level of financing: up to 100% of the total eligible costs

  - call 2/P/2017 "MŁODZI NAUKOWCY 2017"

    - Applicant: research institution/ research consortium/ research- industrial centre/ entrepreneurs—project conducted by research team with young researcher as a Project Manager
    - Project manager: researchers under the age of 38 in the year of submitting project proposal
    - Duration: up to 36 months
    - Level of financing: up to 100% of the total eligible costs but project costs are part of evaluation
    - Budget

      - programme: overall (not just for cybersecurity) 100,000,000 PLN (23,000,000 EUR); 60,000,000 PLN (14,000,000 EUR) in 2016 and 40,000,000 PLN (9,300,000 EUR) in 2018
      - call: not specified
      - project: not specified

    - TRL: VI

  - call 8/2016

    - Applicant: research consortium/ research-industrial centre/ entrepre-neurs
    - Project manager: not specified
    - Duration: not specified
    - Level of financing: up to 100% of the total eligible costs but own contribution and project costs are part of evaluation
    - Budget: not specified
    - TRL: VI

- **CyberSecIdent**—CyberSecurity and e-Identity

  - Applicant: research consortium
  - Project manager: not specified
  - Duration: up to 36 months

---

[3]Example calls: https://www.ncbr.gov.pl/en/programmes/programmes-and-projects-defence-security/.

- Level of financing: up to 100% of the total eligible costs for research institutions, 40–80% of the total eligible costs for entrepreneurs depending on size and kind of work
- Budget

  · programme: not specified
  · call: 1st call—70,000,000 PLN (16,300,000 EUR), 2nd call—31,370,000 PLN (7,300,000 EUR)
  · project: not less than 3,000,000 PLN (700,000 EUR) of the total eligible cost and up to 20,000,000 PLN (4,650,000 EUR) of financing

- TRL: not specified

National Science Centre (NCN)—government agency supervised by the Ministry of Science and Higher Education[4] —calls mainly for researchers, usually about basic research. A few, selected types of competitions are presented below. Most calls under the following types of competitions are announced yearly or twice a year:

- **PRELUDIUM**[5]:

  - Applicant: researchers who do not hold doctorate degrees
  - Project manager: researcher who does not hold a doctorate degree
  - Duration: 12/24/36 months
  - Level of financing: 100%
  - Budget

    · programme: not specified
    · call: 30,000,000 PLN (7,000,000 EUR)
    · project: 70,000 PLN (16,3000 EUR) for projects with duration of 12 months /140,000 PLN (32,6000 EUR) for projects with duration of 24 months/ 210,000 PLN (49,000 EUR) for projects with duration of 36 months

  - TRL: not specified

- **OPUS**[6]:

  - Applicant: research institutions, also consortia
  - Project manager: can indicate between 3 and 10 published research works
  - Duration: 12/24/36 months
  - Level of financing: 100%
  - Budget

    · programme: not specified
    · call: 300,000,000 PLN (70,000,000 EUR)
    · project: not specified

---

[4]https://www.ncn.gov.pl/?language=en.

[5]Example call—PRELUDIUM 17: https://www.ncn.gov.pl/ogloszenia/konkursy/preludium17?language=en.

[6]Example call—OPUS 17: https://www.ncn.gov.pl/ogloszenia/konkursy/opus17?language=en.

– TRL: not specified

- **MINIATURA 2**[7]:

  – Applicant: researchers with a doctoral degree obtained within the last 12 years
    prior to the submission of the proposal
  – Project manager:
  – Duration: 12 months
  – Level of financing: 100%
  – Budget

    · programme: not specified
    · call: 20,000,000 PLN (4,650,000 EUR)
    · project: 5,000–50,000 PLN (1,160–11,600 EUR)

  – TRL: not specified

Grant-in-aid from national ministries (article 37 of the Act on research institutes), e.g., The Ministry of Digital Affairs

- Applicant: research institute
- Project manager: not specified
- Duration: not specified
- Level of financing: 100%
- Budget : not specified
- TRL: not specified

Mixed
In that case, part of the funding comes from the national resources, and part of the
European funds.
Structural measures, i.e. Operational Programmes, e.g., Digital Operation [8]
Operational Programme Digital Poland (OPDP) is funded from two sources: Euro-
pean Regional Development Fund, where the commitment amounts to 2.2 billion
euros, and National funds—public and private, where the minimum commitment
amounts to 394.4 million

- Level of financing: some projects may require beneficiary's own contribution
- Budget for priorities:

  – common access to high-speed Internet: EUR 1 020.2 million,
  – e-government and open government: EUR 949.6 million,
  – digital competences of the society, i.e. encouraging to use the Internet: EUR
    145 million.

---

[7]Example call: https://www.ncn.gov.pl/finansowanie-nauki/konkursy/typy/13?language=en.
[8]https://www.polskacyfrowa.gov.pl/en/site/learn-more-about-the-programme/discover-how-the-
programme-works/funding/.

3.1.2 In Japan

This subsection identifies and describes the mechanisms to finance cybersecurity research in Japan.

- **Ministry of Internal Affairs and Communications (MIC)** is primarily responsible for R&D spending for the communication aspects of cybersecurity, including mobile networks. Historically, MIC initiated some of the joint calls with European Commission on the FP7 and H2020 programme.

  - **Strategic Information and Communications R&D Promotion Programme (SCOPE)** is the research funding instrument of MIC. Normally, annual open calls solicit and adopt proposals based on their quality. Proposals and projects are evaluated by the committee of experts which is usually a mix of industry professionals and academia. Some of the past joint calls with European Commission are operated by building on top of this programme.
  - **National Institute of Information and Communications Technology (NICT)** is the specialized agency under MIC, which operates a limited number of commissioned research in addition to conducting research within its own research centers. Some of the past joint calls with European Commission are operated under the commissioned research programme of NICT. NICT drives its own research centers and allocates most of its resources to these centers, which normally do not initiate calls for proposals.

- **Ministry of Economy, Trade and Industry (METI)** is primarily responsible for R&D spending for the computing aspects of cybersecurity, including industrial control systems.

  - **New Energy and Industrial Technology Development Organization (NEDO)** is the research funding instrument of METI. Normally, annual open calls solicit and adopt proposals based on their quality as well as the impact to the industry.
  - **Information-technology Promotion Agency (IPA)** is one of the specialized agencies under METI, which operates a broad array of programmes for cybersecurity as well as information technology in general. In contrast with other research-oriented specialized agencies, the IPA can be considered to be an operating agency. In order to run its broad array of programmes, IPA closely works with other businesses and research units through open calls, some of which may have elements of research.

- **Ministry of Education, Culture, Sports, Science and Technology (MEXT)** is primarily responsible for funding universities. It initiates calls for transformation of universities, e.g., for internationalization, and for more industry-oriented training programs including cybersecurity. Some of its past funding contributed to improve cybersecurity.

  - **Japan Society for the Promotion of Science (JSPS)** is one of the specialized agencies under MEXT, which operates a broad funding program for academic

research. Its Grants-in-Aid for Scientific Research program (Kakenhi) are competitive funds that are intended to significantly develop all scientific reseach.
JSPS does not operate any funding program specifically for cybersecurity.

- **Cabinet Office (CAO)** is primarily responsible for promoting inter-ministerial
  and inter-sectional R&D efforts through the Cross-ministerial Strategic Innovation
  Promotion Program (SIP). SIP is a national project for science, technology and
  innovation, spearheaded by the Council for Science, Technology and Innovation, in
  order to accomplish its role in leading science, technology and innovation beyond
  the framework of government ministries and traditional disciplines. The SIP has
  identified 10 themes that will address the most important societal problems Japan
  is facing, as well as contribute to the resurgence of the Japanese economy. SIP
  designates Cyber-Security for Critical Infrastructure as one of the 10 themes so that
  research organizations engage in the development of core technologies and social
  implementation technologies under the guidance of the Cybersecurity Promoting
  Committee.

3.1.3 Summary

The following table summarizes the previous section and shows a comparison of the
financing mechanism in the European Union (on the international and national level)
and Japan.

The table shows that in the case of Member States of the European Union, we
have a lot of financing mechanisms that support research and innovation in the
field of cybersecurity. In addition to international funding, there are also national
funding mechanisms that may vary depending on the partner country. However, on
the Japanese side, we have three ministries that finance research and innovation in
the field of cybersecurity and the SIP program (Table 3.1).

Budget

During our research, we tried to obtain information on general funds invested in
cybersecurity research in different ICT areas for both regions. Unfortunately, it seems
that performing such an analysis—either in Japan or EU—is a major task requiring
more resources than available due to the number and complexity of relevant documents and expert decisions involved, such as estimating the investment share in calls
where cybersecurity is only part of the topic. This task involves a review of a very
large number of documents, in many cases involving rough and difficult estimations
(e.g., calls where cybersecurity-related issues are a significant part—but only a part
of the topic). For example, we have analysed "A guide to ICT-related activities in
WP2018-20" which was designed to help potential proposers find ICT-related topics
across the different parts of H2020 in work programme 2018-20. This guide shows
some of the priorities that have been raised, for example, "Boosting the effectiveness
of the Security Union" (where 1.08 billion euros was planned), which includes in

Table 3.1 Comparison of the financing mechanism in the European Union and Japan

European Union	Japan
European Union, International	
• Framework Programmes (now Horizon 2020)—open competitions	
• Special programmes such as Connecting Europe Facility (CEF)—Cybersecurity calls	
• ENISA-European Union Agency for Network and Information Security	
• EUREKA	
National level	
• National programmes	• Ministry of Internal Affairs and Communications (MIC)
• Government, state programmes	• Ministry of Economy, Trade and Industry (METI)
• Regional programmes	• Ministry of Education, Culture, Sports, Science and Technology (MEXT)
• National agencies	• Cabinet Office (CAO)
• International, joint programming	
• Structural measures, i.e. Operational Programmes	
• Funding for research, in general to support Ph.D. grants	
• Other	

particular cybersecurity activities. According to the H2020 calls, it was granted over 508.5 million euros for 2018–2020 (282.5 million in 2018, 179 million in 2019 and 47 million in 2020), but, as shown in Sect. 3.2.2, there are many other calls that are not only concerned with cybersecurity, but are connected. At the moment, some works are in progress. When publishing the cybersecurity package (September 2017), the EC announced more activities in the field of supporting Research and Innovation in cybersecurity. At the same time, work is in progress on a new budget, where dedicated funds for Research and Innovation in cybersecurity are also to be allocated. However, this only refers to the European Commision's funding activity. There are also many financing mechanisms other than that of the European Commission, at both international and national level, as described in previous sections. Finally, there are some major investments that are not part of the regular mechanisms we have identified so far, especially on national and multi-national levels.

According to the "Science and Technology budget overview for 2018" by the Council for Science, Technology and Innovation (CSTI) for year 2018, the planned expenditure on cybersecurity and related fields, that are visible on the CSTI budget overview, were as follows:

- Quantum cryptography for satellite communication: 310 million JPY
- IoT security comprehensive package: 600 million JPY
- AIP: AI/BigData/IoT/Cybersecurity integration project: 8560 million JPY
- Cyberattack countermeasure testbed: 1000 million JPY

This budget overview has significant problem however:

1. Most of the big cybersecurity projects known for the authors, are not included in the overview, each of which exceeds 10 million euros/year;
2. Details of most of the big expenditures are not available, like AIP.

Although AIP's funding is very big (exceeding 80 million euros), its major constituencies are AI, big data and IoT with some links to cybersecurity, thus the precise amount of cybersecurity investment cannot be derived from publicly available sources.

Further evidence can be found in the FP7 JEUPISITE D2.6, "Update of the Inventory of the STI Programmes", which states that *the Japanese STI programmes landscape is rather fragmented as 319 programmes have been identified during a scan carried out in February/March 2016.*

3.2 The Main Research Directions in the Field

A top-down approach is utilized in this section. The analysis of the main research directions in the field is started on the basis of strategic documents regarding cybersecurity, and in particular, Strategic Research and Innovations Agendas, and, on a limited scope, on the basis of national cybersecurity strategies. Although we try to take into account most of the important aspects of cybersecurity, a special emphasis is given to the research and innovation aspect. This analysis is provided in Sect. 3.2.1. Section 3.2.2 is devoted to tactical aspects of implementation of the strategies, as it contains a brief, substantive analysis of programs and project calls regarding cybersecurity. Section 3.2.3 is dedicated to more detailed, and to some extent operational, aspects of cybersecurity, as it includes a brief review of the most important and recent projects in the area of cybersecurity. Section 3.2.4 very briefly describes major research coordination centers and initiatives in the area of cybersecurity, such as European cybersecurity competence centre and network and Digital Innovation Hubs. Section 3.2.5 includes activities other than projects, which are important from the perspective of research and innovation. Each analysis is performed at both the EU and Japan levels. Each analysis focuses on the EU perspective as a whole, but in some specific areas where it was especially relevant, we also considered the perspective of the individual Member States of the EU. The schematic overview of the analysis is presented in Fig. 3.1.

Fig. 3.1 Schematic overview of the analysis

3.2.1 Strategies and Research and Innovation Agendas

This chapter provides an analysis of the "strategic" perspective, which is done on the basis of strategic documents regarding cybersecurity.

3.2.1.1 European Top-Level Strategic Documents

Digital Single Market (DSM) Strategy

The fundamental assumption of the DSM Strategy is to remove regulatory barriers in digital affairs in a way which enables an introduction of a Common European Digital Market. The aim of the document is to facilitate faster development of digital services, and as a result, to improve competitiveness of European companies. The scope of this strategy is broader than only cybersecurity and privacy, but discussion of these aspects is an important part of the document. The document states that only secure digital services can be effectively used by citizens. The main aspect of the document regarding cybersecurity is strengthening the trust in digital services and personal data processing and increasing the security of such services and data. The strategy was published on 6 May 2015 as *Communication from the Commission to the*

European Parliament, the Council, the European Economic and Social Committee and the Committee of the Regions—A Digital Single Market Strategy for Europe [1].

Cybersecurity Strategy of the EU

The European Commission on 7 February 2013 has published the document *Joint Communication to the European Parliament, the Council, the European Economic and Social Committee and the Committee of the Regions—Cybersecurity Strategy of the European Union: An Open, Safe and Secure Cyberspace* [2]. This strategy is historically the first strategic document in the context of cybersecurity and is related to the NIS directive. The strategy indicates five main priorities:

- *Achieving cyber resilience*
- *Drastically reducing cybercrime*
- *Developing cyberdefence policy and capabilities related to the Common Security and Defence Policy (CSDP)*
- *Develop the industrial and technological resources for cybersecurity*
- *Establish a coherent international cyberspace policy for the European Union and promote core EU values*

In many cases, it emphasized that many of these efforts should be supported by research and development and by closer cooperation between governments, especially within the EU. This statement can be of course easily extended to ensure cooperation between the EU and other countries, for example Japan. The Commission stressed out also the importance of cooperation in the area of cybersecurity between public and private sectors. The strategy states that all threats, incidents and crimes, made in cyberspace, cannot be easily affiliated with any borders (for example a specific country) and because of that the cooperation between bodies of information and network security, CSIRTs, telecoms, and so on should be established at the national, EU and also international level.

On 13 September 2017, as a part of the so-called "Cybersecurity Package", the European Commission has published *Joint Communication to the European Parliament and the Council—Resilience, Deterrence and Defence: Building strong cybersecurity for the EU* [3], which is in fact a revision of the Cybersecurity Strategy of the EU. In the document, a strong emphasis is put on common response of Member States in case of international incidents and on cooperation between civil and military sectors. In the opinion of the Commission, cybersecurity should be based on three pillars, and in each pillar, the description of the specific actions was presented in the document:

- Building EU resilience to cyber attacks

  - Strengthening the European Union Agency for Network and Information Security
  - Evolving towards a Single Cybersecurity Market
  - Implementing the Directive on the Security of Network and Information Systems in full
  - Building resilience through rapid emergency response

- Creating a cybersecurity competence network with a European Cybersecurity Research and Competence Centre
- Building a strong EU cyber skills base
- Promoting cyber hygiene and awareness

• Creating effective EU cyber deterrence

- Identifying malicious actors
- Stepping up the law enforcement response
- Encouraging public-private cooperation against cybercrime
- Stepping up the political response
- Building cybersecurity deterrence through the Member States' defence capability

• Strengthening international cooperation on cybersecurity

- Cybersecurity in external relations
- Cybersecurity capacity building
- EU-NATO cooperation.

Another relevant strategic document in the area of cybersecurity is *Communication from the Commission to the European Parliament, the Council, the European Economic and Social Committee and the Committee of the Regions—Strengthening Europe's Cyber Resilience System and Fostering a Competitive and Innovative Cybersecurity Industry* [4], which was released on 5 July 2016. It can be perceived as an update to the Strategy and is strictly connected to the NIS directive (both documents were published around the same time). This document includes declarations of actions which will be taken by the Commission to strengthen the cybersecurity in the EU. The document states that the cooperation in the area of cybersecurity is truly crucial. The announcement of an action plan in the case of international cyber-crisis is also included. For the very first time, the Commission announces actions for cooperation between cybersecurity and crisis management. Very important is the fact that the document announces the creation of a Joint Research Centre to gather expert knowledge, currently dispersed among Member States and union agencies. The Joint Research Centre will be supported by ENISA and CERT-EU. The document also announces the setting up of an advisory group on cybersecurity composed of experts and decision-makers from industry, academia, civil society and other relevant organisations. The group would assist the Commission in preparing new regulations. The document also states that an education platform for cybersecurity will be built up. Promotion of cooperation at the EU level between sectorial entities and evaluation of risk of cyber incidents in similar sectors is also described. One very important conclusion of this document regards the fact that improving the cybersecurity level of small and medium enterprises (SMEs) and improving competences of SMEs as well as various forms of a contractual Public Private Partnership will be supported by the Horizon 2020 programme (Table 3.2).

Table 3.2 Fulfilled objectives by National Cybersecurity Strategies in the EU Member States

Objective	Status	Remarks
Set the vision, scope, objectives and priorities	–/ ✓	Not all of the national strategies provide the strategic vision as well as the main aims with priorities
Follow a national risk assessment approach	–/ ✓	A few strategies state that development and implementation of a risk management system at the national level is one of the most important actions
Take stock of existing policies, regulations and capabilities	–/ ✓	In many strategies, there is no review of the existing regulations regarding cybersecurity and privacy. In many others, only a few of the existing regulations are taken into account
Develop a clear governance structure	✓ / –	Improving the structure of the national cybersecurity system is one of the actions defined by most strategies
Identify and engage stakeholders	✓ / –	see above
Establish trusted information-sharing mechanisms	–/ ✓	Only a few of the strategies state this objective in a direct way
Develop national cyber contingency plans	✓	Most of the strategies satisfy this objective in a direct or indirect way
Protect critical information infrastructure	✓	Most of the strategies state that actions will be taken to improve cybersecurity of critical infrastructure operators
Organise cyber security exercises	✓	Many strategies indicate the need for various types of cybersecurity exercises, such as: comprehensive exercises simulating nationwide incidents, smaller-scale exercises, including sectoral ones, exercises to carry out military operations in cyberspace and also international exercises
Establish baseline security requirements (measures)	✓	Most of the strategies satisfy this objective in a direct or indirect way.
Establish incident reporting mechanisms	✓	Most of the strategies satisfy this objective in a direct or indirect way
Raise user awareness	✓	In most strategies, objectives in the area of raising user awareness are defined
Strengthen training and educational programmes	✓	The need for many actions toward training and education are identified in most of the strategies
Establish an incident response capability	✓	Most of the strategies satisfy this objective in a direct or indirect way
Address cyber crime	✓	Most of the strategies satisfy this objective in a direct or indirect way

(continued)

Table 3.2 (continued)

Objective	Status	Remarks
Engage in international cooperation	✓	In many strategies, various types of cooperation are listed, such as: at the European level, in the NATO and within the United Nations
Establish a public/private partnership	✓/–	The goal of building cooperation mechanisms between the public sector and the private sector is defined in some of the strategies
Balance security with privacy and data protection	✓/–	Most of the strategies satisfy this objective in a direct or indirect way
Institutionalise cooperation between public agencies	✓	In most of the strategies, the objective is met by various proposed actions
Foster R&D	✓	Most of the strategies satisfy this objective in a direct or indirect way
Provide incentives for the private sector to invest in security measures	✓ /–	A few of the strategies meet this objective partially by specific actions (such as creating innovation hubs)

Examples of actions in line with the strategy

Many initiatives and activities can be indicated as part of the implementation of the strategy. The state of adoption of the NIS Directive in particular Member States is advanced. There are also initiatives for broad implementation of the NIS Directive. From the perspective of the EU as a whole, a very important action was to establish a permanent mandate for the European Union Agency for Network and Information Security (ENISA), to replace its limited mandate that would expire in 2020, as well as more resources allocated to the Agency to enable it to fulfil its goals. It is also important to create a framework for European Cybersecurity Certificates for products, processes and services that will be valid across the EU.

3.2.1.2   Strategic Research and Innovation Agenda

The main aim of the Strategic Research and Innovation Agenda [8] is to suggest potential topics for calls of project proposals related to cybersecurity for the H2020 Work Programme 2018–2020. Because of that, the document cannot be perceived as a substitution of the initial cPPP Strategic Research Innovation Agenda (SRIA) v1.0 [9]. The main challenges described in the various ECSO documents regarding research and innovation in cybersecurity, are the following:

- *Market fragmentation;*
- *Innovation led by imported ICT products;*

- *Need to mitigate cyber security dependencies from external sources and achieve strategic supply chain in the field;*
- *Less funding to research and innovation available and often dispersed due to a lack of transnational approach;*
- *European industrial policies not yet properly addressing specific cybersecurity issues;*
- *Weak entrepreneurial culture and lack of venture capital;*
- *Human factor and skills shortage* [8].

All of the identified challenges need very concrete actions based on the harmonization of the European approach to all aspects of cybersecurity. Moreover, such actions should be based on a high-level cybersecurity policy.

The main aims of the outcomes of the Research and Innovation Agenda can be summarized as follows:

- *Protect critical infrastructures and vertical sectors from cyber threats;*
- *Increase European digital autonomy;*
- *Provide security and trust of the whole supply chain;*
- *Invest in areas where Europe has a clear leadership or strategic needs;*
- *Leverage upon the potential of SMEs;*
- *Increase competitiveness* [8].

The most important guidelines agreed in the cPPP SRIA are the following:

- *Allow funding of applied research and innovation, by focusing resources on fields that could maximize European competitiveness in the cyber security market;*
- *Concentrate efforts on cyber security sectors strategic to Europe (and its digital sovereignty);*
- *Foster continuous innovation also through further research for basic technologies and components as well as look forward for possibly disruptive technologies in order to keep medium/long term competitiveness for the products and services* [8].

Overall strategy for the European cybersecurity market and industry

The European cybersecurity market share is about 24%, which is less than the contribution of Europe to Global GDP (i.e. about 26%). It could be perceived as a result of the trends indicated in the previous sections. Many activities should be performed to boost cybersecurity research. For example Strategic Research and Innovation Agenda (SRIA) states that: *"There is an increasing need to reduce the distance with the world-wide research communities dealing with cybersecurity and privacy issues. While there are currently several coordination actions, with US and Japan it would be useful to extend all the international activities to several other countries"* [8]. There is also a plan to identify and carry out many activities with other countries than US and Japan.

Scope

In the following paragraph we briefly describe the main areas identified as promising in the SRIA [8]: *In order to implement the research and innovation strategy and to align technical with cooperation and coordination aspects, five major types of mechanisms/projects are recommended:*

- **Cyber Coordination (Coordination and Support Actions)**: *These projects will foster cooperation (also international) for the efficient sharing of information and coordination of activities.*
- **Cyber Ecosystem**: *Combination of organisational and technical elements—will allow challenges to be addressed in an interdisciplinary way and will serve as a hub for research, innovation, standardisation / certification, experimentation and transfer to market activities.*
- **Cyber Pilots**: *These projects, mainly innovation based, are devoted to piloting solutions in specific vertical domains. These pilots or demonstrators will possibly use the transversal cyber infrastructures and the capabilities developed in the technical projects to demonstrate how the developed innovations can satisfy specific requirements in key vertical sectors, garnering attention and the commitment of users and potential procurement bodies.*
- **Cyber Infrastructures**: *Large (lighthouse) projects that will help to develop large infrastructures in the cyberspace, mainly spanning across several domains with a goal to create a direct competitive advantage to industry and of strategic relevance for European countries. It includes large scale projects which could be funded through a number of different channels, including Horizon 2020 and structural funds. They are specifically designed to raise awareness of the Partnership and give it increased visibility. These large infrastructures need to have a sufficient budget (between 10 and 20 million euros overall) to provide significant results and impact.*
- **Technical projects**: *Small or medium scale technical projects, often R&I activities for developing new cybersecurity capabilities and components. We should ensure that these projects contribute to develop the technical competences and contribute to the KPIs of the cPPP. These projects would be based on clearly defined technical priorities* [8].

In the SRIA [8] seven main priority themes for intervention are outlined:

- *Ecosystem for Education, training, market growth and SME support*

  - *Cyber Range and simulation*
  - *Education and training*
  - *Certification and standardisation*
  - *Dedicated support to SMEs*

- *Demonstrations for the society, economy, industry and vital services*

  - *Industry 4.0*
  - *Energy*
  - *Smart Buildings & Smart Cities*
  - *Transportation*
  - *Healthcare*
  - *E-services for public sector, finance, and telecommunications*

- *Collaborative intelligence to manage cyber threats and risks*

  - *GRC: Security Assessment and Risk Management*
  - *PROTECT: High-assurance prevention and protection*
  - *DETECT: Information Sharing, Security Analytics, and Cyber-threat Detection*
  - *RESPONSE and RECOVERY: Cyber threat management—response and recovery*

- *Remove trust barriers for data-driven applications and services*

  - *Data security and privacy*
  - *ID and Distributed trust management (including DLT)*
  - *User centric security and privacy*

- *Maintain a secure and trusted infrastructure in the long-term*

  - *ICT infrastructure protection*
  - *Quantum resistant crypto*

- *Intelligent approaches to eliminate security vulnerabilities in systems, services and applications*

  - *Trusted supply chain for resilient systems*
  - *Security-and privacy by-design*

- *From security components to security services*

  - *Advanced Security Services*

Examples of Actions in Line with the Strategy

Many activities in line with the strategy can be indicated. As an example, laureates to the H2020 ICT-03 calls have been announced. Four EU pilot projects have been created and are expected to strengthen the EU cybersecurity capacity and tackle future cybersecurity challenges. Initiatives have also been carried out to coordinate efforts in the area of cybersecurity, such as the creation of European cybersecurity competence centre and network, and the Digital Innovation Hub in cybersecurity.

3.2.1.3   Other ECSO WG Publications

The SRIA document, described in the previous section, was published by ECSO WG6. Moreover, other working groups of ECSO also have provided very important publications in the context of cybersecurity strategy for the EU.

Education, awareness, training, cyber ranges

The very important analysis describing gaps in European cyber education and professional training can be found in the ECSO WG5 Position Paper [6]. The document states that there is clearly a lack of qualified cybersecurity professionals and there is a need for a rise in the overall level of cyber awareness. This statement is in line with the SRIA document's findings.

3.2.1.4   National Cybersecurity Strategies in the EU Member States

The analysis of the cybersecurity strategies on national level, both in the EU Member States and in Japan was done on the basis of the methodology provided in the first part of the Appendix. This section only presents the most important findings of the analysis of Member States' cybersecurity strategies, whereas the description of national strategies is provided in the second part of the Appendix. In the next section, a brief description of the cybersecurity strategy of Japan is provided.

It is worth noting that ENISA supports EU Member States in developing, implementing and evaluating their National Cyber Security Strategies (NCSS). Currently, all Member States have developed their NCSS, but the scope of the documents, as well as their actuality, are different.

Strategic Vision

Different countries in different ways define their strategic vision. Most of them put an emphasis on gaining resilience to attacks and threats from cyberspace and creating a safe environment to make use of digital economy with respect to rights and freedoms of citizens.

The Main Aim

The main aim is defined in various ways, but almost every strategy tries to emphasize the importance of cybersecurity. For example in France, cybersecurity is perceived from the military point of view, whereas in other countries, the civil area seems to be more important.

Specific Aims

Specific aims of cybersecurity strategies also vary, but the most common examples of specific aims in the EU Member States are listed below:

- gaining the ability to coordinate actions at the national level, and (in some cases) European or international level,
- gaining the ability to perform cyber threats intelligence and cyber threats prevention,
- fostering research and innovation actions in the area of cybersecurity.

Areas of Interest—Research and Innovation Context

In most of the strategies, the importance of research & innovation and privacy activities is noticed. Some of the strategies define a need for investing in the development of industrial and technological resources for cybersecurity. However, strategies, as very high-level documents, do not provide specific information regarding planned research and innovation actions, they outline main directions of such actions. It is very important that all analysed countries have prepared documents of the strategies although their granularity varies.

The most common areas of research in the field of cybersecurity and privacy which were identified in the strategies are the following:

- Internet of Things,
- Smart Cities,
- Industry 4.0,
- Cloud Computing,
- Big Data,
- methods of attacks and ways of counteracting these attacks,
- assessment of the effectiveness of response to threats,
- cyber risk management at national level.

It is worth noting that similar areas of research were also identified as very promising in questionnaires received during and after the EUNITY workshops. Very similar areas were listed both by European and Japanese responders.

Areas of Interest—Legal, Policy and Organizational Perspective

Most of the strategies state that there is a need to improve organization and coordination of cybersecurity activities at the research level as well as at the operational level. Specifically the most common identified needs are the following:

- The need for legal changes, changes in the structure of cybersecurity system.
- The need for cybersecurity risk management at national level. The need for critical infrastructure, essential services and digital services protection.
- The need to enable military actions in cyberspace.
- The need to build competences in the area of cybersecurity and building awareness in the area of cybersecurity.
- The need to build public-private partnerships, stimulation of research and development.

It is worth noting that detailed regulations are scattered across different legal acts, on various levels, setting the prerogatives of different agencies or regulating the operation of relevant sectors of economy (e.g., energy). The current moves to harmonize and clarify the cybersecurity field (performed both at the level of the EU and Member States), are therefore both necessary and promising.

Areas of Interest—Financial Perspective

In most strategies, there is no specific information regarding funding. However, some point out that specific planning and executive documents will be created. A few strategies also stress the need to prepare a research programme aimed at preparation and implementation of new cybersecurity methods or even the need to create a dedicated institution, which aims at coordinating or even at directly conducting research in the area of cybersecurity and privacy.

Fulfilled Objectives

The description of the objectives and the methodology of analysis are provided in the Appendix.

Examples of Actions in Line with the Strategy

While the state of implementation of the strategy in various states is different, many initiatives have been performed in line with the strategies at the national level. In Poland, for example, the Polish Parliament passed the Act on the National Cybersecurity System, which entered into force on 28 August 2018. The Act implements the provisions of the NIS Directive and introduces a national cybersecurity system. Several research projects conducted in Poland in the area of cybersecurity and privacy are in line with the strategy and also with the Act. National Cybersecurity Platform is a very large research and development project whose aim is to create an interactive system of monitoring and visualization of the actual state of national cyberspace security. The system will use various methods of dynamic and static risk analysis, an expert system for decision support as well as many tools for vulnerabilities and threats detection.

3.2.1.5 National Strategy in Japan

Brief Description

In Japan, the Basic Act on Cybersecurity was adopted in November 2014 in order to designate responsibilities and authorities to specialized agencies. The National Cybersecurity Strategy has been initially adopted by the cabinet in September 2015 and subsequently revised in July 2018, which also predicates annual update of the cybersecurity planning by the Cybersecurity Strategic Headquarter. The Government of Japan has laid out this strategy as well as the annual plan as the platform for the common understanding and actions of relevant stakeholders. Japan has specific law regulating a certain industry, where mandatory standards and guidelines, including the mandatory reporting of cybersecurity incidents, are set forth by competent ministries for the purpose of critical infrastructures protection. Its National Cybersecurity Strategy clearly identifies the scope, whose details are further documented in the Cybersecurity Policy for Critical Information Infrastructure Protection [11]. (Table 3.3).

Previous Strategies and Related Documents

Previous strategies were published by the Information Security Policy Council, which has been transformed into the Cybersecurity Strategic Headquarter of the Cabinet in accordance with the Basic Act of Cybersecurity in 2015. Since 2015, the Cybersecurity Strategy has been augmented by the annual plan issued by the headquarter. They are respectively available as Cybersecurity 2015 through 2019.

National R&D strategy was initially adopted in July 2011 by the Information Security Policy Council, based on the recommendation of the 4th Science and Tech-

Table 3.3 Basic information about the National Cybersecurity Strategy in Japan

Owner (responsible entity)	Government of Japan
Level of public administration—acceptance/publication	Government
Date of publication	4 September 2015 (revised 27 July 2018)
Type of document	Strategy
English name	Cybersecurity Strategy
Language versions	Japanese[a] and English[b]
Status	In force
Scope (geographical or sectoral)	Japan
Scope (on the basis of merit)	All aspects related to cybersecurity
Types of stakeholders	All

[a] Available at: https://www.nisc.go.jp/active/kihon/pdf/cs-senryaku-kakugikettei.pdf and https://www.nisc.go.jp/active/kihon/pdf/cs-senryaku2018.pdf
[a] Available at: https://www.nisc.go.jp/eng/pdf/cs-strategy-en.pdf and https://www.nisc.go.jp/eng/pdf/cs-senryaku2018-en.pdf

nology Basic Plan. The R&D strategy was subsequently revised in July 2014. The strategy reflected R&D activities of competent ministries and specialized agencies.

The latest R&D strategy was published by the Cybersecurity Strategic Headquarter as the "Cybersecurity Research and Development Strategy" in July 2017. The strategy can be summarized by the following 7 points:

1. Consideration of business process in the cybersecurity R&D
2. Security consideration across the life-cycle of products and services
3. Holistic approach, including technology, legislation, psychology and management
4. Integration of cyberspace and physical space through IoT
5. Use of artificial intelligence and big data
6. Public-private partnership and the industry-academic collaboration
7. Facilitation of cybersecurity R&D through legal measures

Strategic Vision

As Japan strives to become the world's highest level IT-based society, reinforcing cybersecurity will be imperative not only for national security and crisis management, but also for bolstering Japan's industrial competitiveness through the use of IT and data.

The Main Aim

The Cybersecurity Strategy identifies the following objective:

Ensure a free, fair, and secure cyberspace; and subsequently contribute to improving socio-economic vitality and sustainable development, building a society where the people can live safe and secure lives, and ensuring peace and stability of the international community and national security.

Japan affirms the following basic principles in policy planning and implementation for reaching the objective: assurance of the free flow of information, the rule of law, openness, autonomy, and collaboration among multiple stakeholders.

Specific Aims

The Cybersecurity Strategy outlines four detailed aims:

1. Improvement of Socio-Economic Vitality and Sustainable Development
2. Building a Safe and Secure Society for the People
3. Contribution to the Peace and Stability of the International Community and Japan's National Security
4. Cross-Cutting Approaches to Cybersecurity

Areas of Interest—Research and Innovation Context

The strategy envisages the cyber-physical space in addition to cyberspace as follows: *All kinds of physical objects and people have been interconnected by ICTs in a more multidimensional way, and the integration of physical space and cyberspace has become more intertwined. Attention should be paid to the fact that any event in cyberspace may affect society as a whole, producing a synergy effect with various events including those in physical space. Recognizing the transformational process leading to the unprecedented interconnected and converged information society, Japan will implement policies by precisely capturing such social transformation.*

The 5th Science and Technology Basic Plan (adopted by the Cabinet on January 22, 2016) incorporated this recognition and coined the term Society 5.0, where smart energy, transport, manufacturing, services will be orchestrated in near future with smart management services such as HR, accounting, and legal services, as well as smart workforce such as the on-demand workforce and innovation, eventually leading to creation of new values. The plan specifically recognizes cybersecurity as the key enabler toward value creation and international competitiveness.

Areas of interest—legal, policy and organizational perspective

The legal basis of cybersecurity can be found in the Article 22 of the Basic Act on the Formation of an Advanced Information and Telecommunications Network Society, which was enacted in December 2000. The Article 22 specifically requires that "Strategies developed to form an advanced information and telecommunications network society shall ensure that action is taken to achieve and maintain security and reliability of advanced information and telecommunications networks, to protect personal information and other actions necessary to enable citizens to use such networks without anxiety", upon which basic principles for cybersecurity were later stipulated in the Basic Act on Cybersecurity. Diverse programs have been implemented by the national government and local governments based on the Basic Act on Cybersecurity.

As cybersecurity measures are implemented by the balancing act of "freedom of expression" and "secrecy of communication", the case of Japan can be an interesting example, as its Constitution mandates *the secrecy of any means of communication* in the Article 21 and further elaborated in the Radio Act, Wire Telecommunications Act, and Telecommunications Business Act.

It is worth noting that laws in Japan define minimum requirements, upon which associated guidelines are defined and published, in order to promote voluntary programs by both public and private entities. As cybersecurity measures effectively require public-private cooperation, the guideline-based approach of Japan can be useful in other countries.

NISC, directly attached to the cabinet secretariat, oversees and coordinates cybersecurity policy of all ministries. It also serves as the headquarter of cybersecurity policy, thus promoting integrated and strategic cybersecurity policy. Historically, NISC was founded in April 2005 as the National Information Security Center, subsequently renamed as the National center of Incident readiness and Strategy for Cybersecurity after the Basic Act on Cybersecurity was enacted. The Act assigns authority to NISC, enabling it to audit and give recommendations to government agencies.

NISC also promotes international cooperation. International Strategy on Cybersecurity Cooperation, which was published in 2013 by the Information Security Policy Council, outlines basic policies for international cooperation as well as priority areas. The document also puts focus on regional cooperation with ASEAN, as well as cooperation with EU, US and multilateral frameworks.

NISC has also been promoting public-private partnership. For instance, the Cybersecurity Policy for Critical Information Infrastructure Protection outlines information sharing to/from NISC through the CEPTOAR council (CEPTOAR: Capability for Engineering of Protection, Technical Operation, Analysis and Response), which exist for 14 critical infrastructure sectors.

Areas of Interest—Financial Perspective

The Japanese government's cybersecurity spending for 2018 was approximately 62 billion yen, which translates to roughly 485 million euros. The expenditure includes cybersecurity measures, basic R&D, and training programs. Current priorities on the budget are: (1) IoT security, (2) critical infrastructure protection and the protection of government agencies, (3) investment into national R&D agencies, (4) capacity building and R&D, and (5) investments toward the Tokyo 2020 Olympic and Paralympic games. As the cybersecurity investments are closely linked with the IT investments, coordinations among them are considered important.

As cybersecurity measures cannot be implemented by the government alone, it is considered important to promote private investments and engagements. In order to promote cybersecurity activities across the private sector, a variety of measures are put in place, including issuance of guidelines and budget allocations.

According to the JNSA IT Security Market Analysis Report 2016, IT security market in Japan was approximately 979 billion yen in 2017, which translates to 7.6 billion euros. It is anticipated to exceed 1 trillion yen in 2018.

Fulfilled Objectives See Table 3.4.

Examples of Actions in Line with the Strategy

The annual plan elaborates on actions in line with the strategy, which is issued by the Cybersecurity Strategic Headquarter as Cybersecurity 2015 through 2019. The annual plan designates competent ministries to work on specific actions in line with the strategy.

Examples of actions are as follows:

CIIREX (Critical Infrastructure Incident Response Exercise) is a cross-sector cyber security exercise which is coordinated by NISC. The exercise is conducted on an annual basis where more than 3000 participants from 14 critical infrastructure sectors evaluate their readiness to cyber incidents through a series of desktop exercises.

CYDER (CYber Defense Exercise with Recurrence) is the training program operated by NICT where participants experience detection, response, reporting and recovery from cyber incidents in a testbed setup emulating organization LAN. Targeted participants are the employees of government ministries, local public entities, specialized agencies and critical infrastructure operators. The exercise has been conducted 100 times annually across major cities in Japan, gathering 3000 participants in total.

3.2.2 Programs and Project Calls

In this section, we provide an analysis of the "tactical" perspective.

3.2.2.1 European Project Calls

To implement the R&D and solutions to cope with the issues on cybersecurity and promote cooperation, the Horizon 2020 is the latest initiative of EU Research and Innovation programme with nearly 80 billion euros of funding available over 7 years (2014–2020)—in addition to the private investment that this money will attract. In the Horizon 2020 2014–2015 ICT Work Programme and 2016–2017 ICT Work Programme, there were a lot of research sectors which were prioritized to push the research and innovation. Those priorities are also present in the ICT Work Programme for 2018–2020. Some of them directly involve cybersecurity and some of them include the cybersecurity aspect in an indirect way. A brief summary of those calls is presented in the next paragraphs.

Information and Communication Technologies (ICT)
Cybersecurity can appear in almost every sub-item of ICT. Here we mention some of them. In the Horizon 2020 2014–2015 ICT Work Programme there were for example:

Table 3.4 Fulfilled objectives by Japanese national cybersecurity strategy

Objective	Status	Remarks
Set the vision, scope, objectives and priorities	✓	Basic Act on Cybersecurity
Follow a national risk assessment approach	✓	
Take stock of existing policies, regulations and capabilities	✓	Review the cybersecurity strategy every three years.
Develop a clear governance structure	✓	Basic Act on Cybersecurity assigns responsibilities to specialized agencies.
Identify and engage stakeholders	✓	Basic Act on Cybersecurity, Article 4–9
Establish trusted information-sharing mechanisms	✓	Sector-specific information-sharing mechanisms, as well as cross-sector information sharing mechanisms exist.
Develop national cyber contingency plans	✓	Cybersecurity Strategy, the Cybersecurity Policy for Critical Infrastructure Protection (4th Edition) etc.
Protect critical information infrastructure	✓	14 critical infrastructure sectors have been identified and are closely monitored and assisted by NISC.
Organise cyber security exercises	✓	Sector-specific exercises and cross-sector exercises are conducted on a regular basis.
Establish baseline security requirements (measures)	✓	Public sector adopts the common security criteria for the government. Private sectors adopt their own baseline security requirements with assistance and oversight of competent ministries.
Establish incident reporting mechanisms	✓	Basic Act on Cybersecurity
Raise user awareness	✓	Awareness programme has been organized by NISC and IPA.
Strengthen training and educational programmes	✓	Educational programmes are managed by MEXT. Training programs for both private and public sectors are managed by specialized agencies
Establish an incident response capability	✓	GSOC (Government Security Operation Coordination Team), JPCERT/CC, and private-sector CSIRTs
Address cyber crime	✓	National Policy Agency works with specialized agencies, private sector entities as well as the Japan Cybercrime Control Center to address cyber crime
Engage in international cooperation	✓	Agencies actively promote international cooperation, as documented in the International Strategy on Cybersecurity Cooperation

(continued)

Table 3.4 (continued)

Objective	Status	Remarks
Establish a public/private partnership	-	Keidanren (Japan Business Federation) engages in cybersecurity dialogue with the government
Balance security with privacy and data protection	✓	The Act on Personal Information and the Basic Act on Cybersecurity outlines the basic framework. Guidelines by competent ministries augment the framework as necessary, which can be used to maintain the balance
Institutionalise cooperation between public agencies	✓	Basic Act on Cybersecurity
Foster R&D	✓	The 5th Science and Technology Basic Plan identifies cybersecurity as one of the four prioritized policy agenda
Provide incentives for the private sector to invest in security measures	✓	Regulation and support from competent ministries, specialized agencies as well as NISC

- ICT 32-2014: Cyber-security, Trustworthy ICT
- ICT 38-2015: International partnership building and support to dialogues with high income countries

In the Horizon 2020 2016–2017 ICT Work Programme there were for example:

- ICT-05-2017: Customised and low energy computing
- ICT-06-2016: Cloud Computing
- ICT-14-2016-2017: Big Data PPP: cross-sectorial and cross-lingual data integration and experimentation
- ICT-35-2016: Enabling responsible ICT-related research and innovation

In the Horizon 2020 2018–2020 ICT Work Programme there were three calls connected with cybersecurity: *Information and Communication Technologies, Digitising and transforming European industry and services: digital innovation hubs and platform* and *Cybersecurity*. The example of topics in that calls are:

Information and Communication Technologies

- ICT-08-2019: Security and resilience for collaborative manufacturing environments (Joint with FoF)
- ICT-10-2019-2020: Robotics Core Technology
- ICT-11-2018-2019: HPC and Big Data enabled Large-scale Test-beds and Applications
- ICT-15-2019-2020: Cloud Computing
- ICT-20-2019-2020: 5G Long Term Evolution
- ICT-27-2018-2020: Internet of Things
- ICT-28-2018: Future Hyper-connected Sociality

Digitising and transforming European industry and services: digital innovation hubs and platform

- DT-ICT-01-2019: Smart Anything Everywhere
- DT-ICT-06-2018: Coordination and Support Activities for Digital Innovation Hub network
- DT-ICT-08-2019: Agricultural digital integration platforms

Cybersecurity

- SU-ICT-01-2018: Dynamic countering of cyber-attacks
- SU-ICT-02-2020: Building blocks for resilience in evolving ICT systems
- SU-ICT-03-2018: Establishing and operating a pilot for a Cybersecurity Competence Network to develop and implement a common Cybersecurity Research & Innovation Roadmap
- SU-ICT-04-2019: Quantum Key Distribution testbed

Only some calls are listed above. There are many others that deal with the cybersecurity aspect, but this is not their main goal.

Joint Collaborations
Several of the Joint calls in International Collaboration are also interested in cybersecurity, such as

- EUB1-2015: Cloud Computing, including security aspects
- EUB2-2015: High Performance Computing (HPC), the priorities focus on the security aspects of all applications of HPC on societal challenges and in areas such as transport, energy, environment, climate, health, etc.
- EUB3-2015: Experimental Platforms, deals with the security related to the current tools and platforms in support of end-to-end experimentation.
- EUB-01-2017: Cloud Computing
- EUB-02-2017: IoT Pilots
- EUB-03-2017: 5G Networks

The other noticeable joint collaborations, namely "EU-Japan Research and Development Cooperation in Net Futures", identified by H2020-EUJ-2014, also mention many cybersecurity issues. The 4 following topics consider security issues as top priorities:

- EUJ1-2014: Technologies combining big data, internet of things in the cloud, creates new challenges, especially for services and data hosted and executed across borders
- EUJ2-2014: Optical communications, cybersecurity should also be addressed and go side by side with the evolution of the network traffic speed and growth to enable the identification of potential threats
- EUJ3-2014: Access networks for densely located users, since those networks should be appropriately secured in order for the citizens to be able to access the provided services

- EUJ4-2014: Experimentation and development on federated Japan-EU testbeds, the proposals should also investigate the related interoperability, privacy and security issues

There is also one topic in which cybersecurity is a key issue in the Horizon 2020 2016–2017 ICT Work Programme under the "EU-South Korea Joint Call" and one in the Horizon 2020 2018–2020 ICT Work Programme under the "EU-Japan Joint Call"

- EUK-01-2016: 5G—Next Generation Communication Networks
- EUJ-01-2018: Advanced technologies (Security/Cloud/IoT/BigData) for a hyper-connected society in the context of Smart City

Innovation In SMEs (INNOSUP)

In the topic of "INNOSUP6-2015: Capitalising the full potential of online-collaboration for SME innovation support", there are obviously security issues for online applications and services.

- INSO-9-2015: Innovative mobile e-government applications by SMEs
- INSO-10-2015: SME business model innovation
- SMEInst-01-2016-2017: Open Disruptive Innovation Scheme
- SMEInst-06-2016-2017: Accelerating market introduction of ICT solutions for Health, Well-Being and Ageing Well
- SMEInst-13-2016-2017: Engaging SMEs in security research and development
- SC1-PM-18-2016: Big Data supporting Public Health policies

Secure Societies—Protecting Freedom And Security Of Europe And Its Citizens (DRS)
There are a few calls where cybersecurity is addressed: *Disaster-resilience: safeguarding and securing society, including adapting to climate change (DRS), Fight against crime and Terrorism (FCT), Digital Security: Cybersecurity, Privacy and Trust (DS)*. The cybersecurity issues are directly addressed within this call:

- FCT-1-2015: Forensics topic 1: Tools and infrastructure for the extraction, fusion, exchange and analysis of big data including cyber-offenses generated data for forensic investigation
- FCT-3-2015: Forensics topic 3: Mobile, remotely controlled technologies to examine a crime scene in case of an accident or a terrorist attack involving CBRNE materials
- DRS-17-2014/2015: Critical infrastructure protection topic 7: SME instrument topic: "Protection of urban soft targets and urban critical infrastructures"

There are also many digital security topics that deal with cybersecurity and privacy (under the Secure Society priority) and we include them in the section below.

- Work Programme 2014–2015

  – DS-1-2014: Privacy
  – DS-2-2014: Access Control

- DS-3-2015: The role of ICT in Critical Infrastructure Protection
- DS-4-2015: Information driven Cyber-security Management
- DS-5-2015: Trust e-Services
- DS-6-2014: Risk management and assurance models
- DS-7-2015: Value-sensitive technological innovation in Cybersecurity

- Work Programme 2016–2017

  - DS-02-2016: Cyber Security for SMEs, local public administration and Individuals
  - DS-03-2016: Increasing digital security of health related data on a systemic level
  - DS-04-2016: Economics of Cybersecurity
  - DS-05-2016: EU Cooperation and International Dialogues in Cybersecurity and Privacy Research and Innovation
  - DS-06-2017: Cybersecurity PPP—Cryptography
  - DS-07-2017: Cybersecurity PPP—Addressing Advanced Cyber Security Threats and Threat Actors
  - DS-08-2017: Cybersecurity PPP—Privacy, Data Protection, Digital Identities

Under the Work Programme 2016–2017 the "CRITICAL INFRASTRUCTURE PROTECTION H2020-CIP-2016-2017" focuses on critical infrastructure protection. An example of a topic is the following:

- CIP-01-2016-2017: Prevention, detection, response and mitigation of the combination of physical and cyber threats to the critical infrastructure of Europe

Under the Work Programme 2018–2020 the "Secure societies—Protecting freedom and security of Europe and its citizens" focuses on protecting the infrastructure of Europe and the people in the European smart cities, disaster-resilient societies, fight against crime and terrorism, border and external security and digital security. The specific topics for that priority are:

- SU-INFRA01-2018-2019-2020: Prevention, detection, response and mitigation of combined physical and cyber threats to critical infrastructure in Europe
- SU-INFRA02-2019: Security for smart and safe cities, including for public spaces
- SU-DRS02-2018-2019-2020: Technologies for first responders
- SU-FCT02-2018-2019-2020: Technologies to enhance the fight against crime and terrorism
- SU-FCT03-2018-2019-2020: Information and data stream management to fight against (cyber)crime and terrorism
- SU-BES02-2018-2019-2020: Technologies to enhance border and external security
- SU-BES03-2018-2019-2020: Demonstration of applied solutions to enhance border and external security
- SU-DS01-2018: Cybersecurity preparedness—cyber range, simulation and economics
- SU-DS02-2020: Management of cyber-attacks and other risks

- SU-DS03-2019-2020: Digital Security and privacy for citizens and Small and Medium Enterprises and Micro Enterprises
- SU-DS04-2018-2020: Cybersecurity in the Electrical Power and Energy System (EPES): an armour against cyber and privacy attacks and data breaches
- SU-DS05-2018-2019: Digital security, privacy, data protection and accountability in critical sectors.

Under the Work Programme 2018–2020 the "Trusted digital solutions and Cybersecurity in Health and Care" focus area on "Boosting the effectiveness of the Security Union" had two topics connected with cybersecurity:

- SU-TDS-02-2018: Toolkit for assessing and reducing cyber risks in hospitals and care centres to protect privacy/data/infrastructures
- SU-TDS-03-2018: Raising awareness and developing training schemes on cybersecurity in hospitals

Inclusion of cybersecurity aspects is now required in many calls from various fields (not directly related to cybersecurity in itself), as for example:

- Future And Emerging Technologies (FET)—Cyber-security as an application field needs a lot of enabling technologies. The Future and Emerging Technologies (FET) programme has three complementary lines of action to address different methodologies and scales, from new ideas to long-term challenges. These are FET Open, FET Proactive and FET Flagships.
- European Research Infrastructures, Including E-Infrastructures (EINFRA)— cybersecurity is directly addressed in e-infrastructures.
- Societal Challenges—Secure, Clean And Efficient Energy (LCE)
- Smart, Green And Integrated Transport (MG)
- Marie Skłodowska-Curie actions (Individual Fellowships (IF): support for experienced researchers undertaking mobility between countries, optionally to the non-academic sector; Research networks (ITN): support for Innovative Training Networks)
- Climate Action, Environment, Resource Effciency And Raw Materials (SC)

Brief Summary

It is worth noting that there is an increase in the total number of calls in the area of cybersecurity. Cybersecurity does not only occur in project calls devoted to this issue but also in project calls from other areas in which cybersecurity is an important element. Inclusion of cybersecurity aspects is now required in many calls from various fields.

3.2.2.2 Examples of National Project Calls

National Defence and Security Calls (Poland)

Calls which regard national security and defence (in very wide scope—from weapons and other military applications to cybersecurity of critical infrastructure). Solutions in the area of cyber defence shall develop abilities of the Armed Forces of the Republic of Poland to provide cybersecurity of ICT and communication systems of the Ministry of National Defence and their ability to operate in cyberspace. These solutions may also be useful in systems connected with the security of critical infrastructures of the Republic of Poland, the crisis management systems and ICT systems used in Police, Border Guards, State Fire Service.

CyberSecIdent (Poland)

Cyber Security and e-Identity. The main goal of the CyberSecIdent programme is to increase security level of cyberspace of the Republic of Poland through the improvement of accessibility of software and hardware tools till 2023. The specific goal is to implement technological solutions to facilitate cooperation and coordination of activities in the area of security of cyberspace with special regard to e-Identity. The other specific goal is to implement methods and techniques of identification and authentication.

NCN (Poland)

Calls support basic research in Poland. Examples of calls:

- **PRELUDIUM** is a funding opportunity intended for pre-doctoral researchers about to embark on their scientific career
- **OPUS** is a funding opportunity intended for a wide range of applicants. The research proposal submitted under this scheme may include the purchase or construction of research equipment
- **MINIATURA** is a scheme offering funding for single activities that serve as part of larger basic research

Grant-in-aid (Poland)

From national ministries (article 37 of the Act on research institute)—research institutes may be assigned tasks by their supervisory ministry.

Operational Programme Digital Poland

Focuses on Internet access, e-government and general digital competence of the society.

3.2.3 Projects

In this chapter there is an analysis of the "operational" perspective provided.

3.2.3.1 National and International Roadmapping Projects

The European Commission has led many coordination and roadmapping projects. The following ones have been identified as relevant. These projects are included in this section but it should be noted that they have very serious strategic meaning—to some extent they create a strategic plan for future activities in the area of cybersecurity and privacy (also in the research and innovation field). The most important projects in this context are the following:

- **CyberROAD, CAMINO, COuRAGE**
  Creating a research roadmap for a topic as dynamically changing as cybersecurity requires great predictive ability. This difficulty was recognized by the European Commission through decision to fund not one, but three separate CSA-type (Coordination and Support Action) projects in the FP7 call SEC-2013.2.5-1—Developing a Cyber crime and cyber terrorism research agenda. The projects tasked with creating the roadmap were (a) CyberROAD—Development of the CYBER crime and CYBER terrorism research ROADmap (project no. 607642), (b) CAMINO—Comprehensive Approach to cyber roadMap coordINation and develOpment (project no. 607406) and (c) COuRAGE—Cybercrime and cyberterrOrism (E)Uropean Research AGEnda (project no. 607949). The three projects took different approaches to the development of the research agenda, from the information collection phase to the form of the final results. CyberROAD provides an interactive roadmap on its website, CAMINO summarizes its conclusions in a book while COuRAGE delivers a set of reports, concluding the work with the COuRAGE Brief. However, while the approaches were very different, the crucial results—the identification and prioritization of research topics—produced by all three projects were similar, giving the hope that the common findings of separate teams of experts in the field will likely be reliable predictions. The results of the three projects were compared and discussed during the joint conference on "Emerging and Current Challenges in Cybercrime and Cyberterrorism", which took place in The Hague, Netherlands, in March 2016, and in a collective book, published by Springer [10]. The results of the projects are likely to influence the future European research agenda in the field of cybersecurity [11].
  It is worth noting that CyberROAD project priorities focus on the identification of cybercrime and cyberterrorism threats and the constitution of a strategic roadmap to handle them. The key directions that CyberROAD follows are built around the technological, societal, legal, ethical, and political aspects of cyber-crime and cyber-terrorism.
- **SECCORD**
  Project SECCORD (Security and Trust Coordination and Enhanced Collaboration) was providing coordination and services for the Trust and Security (T&S) research programme and its projects. One of the goals of the project is to map research results of FP7 security research projects to the ongoing ICT security industry (supply-side) priorities and offerings, market trends and demand-side needs and user priorities. One of the main objectives was to reduce the gap in the alignment of research to demand-side needs, besides supply-side priorities (often expressed in industrial research agendas) or market trends (found in different market analyst reports, hype curves, emerging technology radars etc.) [11].

- **INCO-Trust and BIC**

  INCO-Trust and BIC (Building International Cooperation for Trustworthy ICT) were two coordination and support actions targeting expansion of co-operation between EU researchers and programme management with their peers in non-EU countries. While INCO-Trust focus was on highly developed countries, including Japan, BIC focus was more on ICT high-growth countries, specifically Brazil, India and South Africa, who represent emergent world-impacting information economies through the scale and sophistication of their growing ICT sectors [11].

- **CIRRUS**

  (Certification, InteRnationalisation and standaRdization in cloUd Security) *was bringing together representatives of industry organizations, law enforcement agencies, cloud services providers, standard and certification services organizations, cloud consumers, auditors, data protection authorities, policy makers, software component industry etc. with strong interests in security and privacy issues in cloud computing* [11].

- **CAPITAL**

  (Cybersecurity research Agenda for PrIvacy and Technology challenges) *was another CSA project that delivered Cybersecurity Research Agenda which provided 34 research clusters, defined areas of research critical for a EU Cybersecurity Strategy, and performed a rating of research clusters regarding impact, expected maturity and timeline* [11].

- **SYSSEC**

  The SysSec Network of Excellence has produced the "Red Book of Cybersecurity" to serve as a Roadmap in the area of Systems Security. SysSec Red Book was realized by putting together a "Task Force" of top-level young researchers in the area steered by the advice of SysSec Work Package Leaders. The Task Force had vibrant consultations (i) with theWorking Groups of SysSec, (ii) with the Associated members of SysSec, and (iii) with the broader Systems Security Community. Their feedback was captured in an on-line questionnaire and in forward-looking "what if" questions. The Task Force was then able to distill their knowledge, their concerns, and their vision for the future. The result of this consultation has been captured in the so-called Red Book which serves as a Road Map of Systems Security Research and as an advisory document for policy makers and researchers who would like to have an impact on the Security of the Future Internet. According to this target the Red Book was developed with specific chapters targeted for specific audiences [11] (Fig. 3.2).

  SysSec intended to initiate a more proactive approach in the way we stand towards cyber attacks and lead to the prediction of the threats and vulnerabilities rather than try to tackle the security issues after an attack. Towards this direction, SysSec targeted the advance of cyber defense by joining hands with Industry and Policy Makers while taking into consideration FORWARD initiative results. The main goals included the constitution of a European Network of Excellence that centralized the Systems Security research; the support of cybersecurity education; the introduction of a think tank to explore current threats and vulnerabilities and the initiation of an effective research direction in cybersecurity. Finally SysSec proposed to estab-

Fig. 3.2 The Red Book.
Produced by the SYSSEC
project (2010–2014) (http://
www.red-book.eu)

lish a collaborative working plan to promote synergetic research. Towards these directions, SysSec organized several workshops and summer schools, provided student scholarships, covering living and travelling expenses for visiting organizations in the SysSec Network of Excellence (including FORTH-ICS (GR), Vrije Universiteit Amsterdam (NL), Institut Eurecom (FR), IPP—Bulgarian Academy of Sciences (BG), TU Vienna (AT), Chalmers University (SE) and Politecnico di Milano (IT)), and created several courses covering Computer Security, Cryptography, Language-based Security, Network Security, Secure Programming, Algorithms and Architectures for Cryptographic Systems, Privacy and Security Engineering.

- **FORWARD**

The main goal of the FORWARD project was to identify future attacks that threaten the security of the ICT infrastructure of Europe. It is clear that ICT infrastructures are no longer independent of societies and the economy, but they have evolved into the enabling and carrying platform that facilitates communication, production, trade, and transportation. Thus, attacks against ICT infrastructures have significant, adverse impact on the well-being of Europe and its people, and it is clear that adequate protection must be provided. By identifying future problems, Europe can take an active approach in defending against anticipated threats. In particular, one can prepare by setting a research agenda that provides solutions before threats become unmanageable. Main target of the project is to collaborate with international experts from academia and industry (both from Europe and the rest of the world) to identify future threats, and find those that are most pressing and that require the most immediate attention [11] (Fig. 3.3).

FORWARD main goals included the identification, networking and coordination of the research efforts in Academia and Industry in order to establish secure and trusted ICT systems and infrastructures. Towards this direction, three working groups discussing the best practices, progress and priorities in cybersecurity,

Fig. 3.3 The FORWARD blog. Produced by the FORWARD project (2008–2009) (http://www.ict-forward.eu/)

were established: Smart Environments (WG-SE) , Malware & Fraud (WG-MF) , and Critical Systems (WG-CS) . FORWARD created an online framework hosting the latest threat detection and prevention mechanisms. Additionally, industry, academia and policy makers organized workshops to explain the emerging threat landscape. The results of the project are presented in the FORWARD Whitebook .

- **CONNECT2SEA**

  In the framework of the CONNECT2SEA project, a report was compiled on common priorities to promote cooperation regarding cybersecurity between Europe and ASEAN countries. The sources used to compile this report are, on the European side, the H2020 work programmes on cybersecurity, Information and Communications Technology (ICT), and data information that has been provided by CONNECT2SEA partners from ASEAN countries. The report determined the top 12 R&D issues which are of the highest interest in ASEAN countries, includ-

ing new advanced R&D trends, such as Cloud Computing, Big Data, Internet of Things, Cryptography, e-Services. The report also presented the results from the two (2) Cybersecurity related workshops that were organized to gain feedback from experts in the area, in particular regarding weaknesses in networking and community cooperation to cope with cybersecurity in ASEAN [11] (Fig. 3.4).

CONNECT2SEA as a follow up project to the SEACOOP, in a 30-months period, established a strategic synergy between European Union and South East Asia in the Information and Communication Technologies (ICT) research. Among the achievements of the project, the transfer of knowledge and the policy dialogue, between the two regions, were set at a high priority. CONNECT2SEA's goals included the advancement of ICT R&D&I partnership, through the internationalization of EU and SEA research and innovation networks and frameworks, the promotion of Horizon 2020 and cooperative research programmes between EU and ASEAN, and the augmentation of EU-SEA ICT R&D policy dialogue. These goals have been accomplished through conferences,[9] grants,[10] funding fellowships[11] and meetings focused on cybersecurity.[12] The main outcomes of the project were the establishment of a common agenda and strategic cooperation between Europe and SEA countries, and the formulation of a knowledge network between the two regions.

- **ECRYPT-CSA**

  This CSA brings together the European cryptography community, involving academia, industry, government stakeholders and defence agencies. The project is coordinating ongoing research, develops a joint research agenda and foresight study, and identifies technology gaps and market and innovation opportunities and coordinate and strengthen standardization efforts; it also addresses governance of cryptographic standards at a European level. The project tackles through advanced training initiatives the skill shortage of academia and industry. The project intends to bridge the gap between academic research on the one hand and standards and industry innovations on the other hand, hereby strengthening the European industrial landscape in a strategic area. The project results in the availability of more trustworthy security and privacy solutions "made in Europe", resulting in an increased user trust in ICT and online services and empowerment of users to take

---

[9]More information available at: http://www.connect2sea.eu/news-and-events/news/details/Towards-new-avenues-in-EU-ASEAN-ICT-collaboration-Highlights-and-presentations.html.

[10]More information available at: http://www.connect2sea.eu/news-and-events/news/details/Newton-Institutional-inks-Grants-to-Develop-Research-and-Innovation-Collaborations.html.

[11]More information available at: http://www.connect2sea.eu/news-and-events/news/details/Humboldt-Research-Fellowship-for-Postdoctoral-Researchers-in-Germany.html,
http://www.connect2sea.eu/news-and-events/news/details/France-15-Postdoc-Positions-in-Biology-IT-at-CEA.html, http://www.connect2sea.eu/news-and-events/news/details/Lise-Meitner-Programme-in-Austrian-Research-Institute.html, http://www.connect2sea.eu/news-and-events/news/details/EURAXESS-ASEAN-Funding-and-Fellowship-in-Hanoi-and-Bangkok.html.

[12]More information available at: http://www.connect2sea.eu/news-and-events/news/details/CONNECT2SEA-Activities-During-APAN-Meetings-Manila-2016.html, http://www.connect2sea.eu/news-and-events/news/details/EU-ASEAN-CIO-Forum-Jakarta-Building-Digital-Platform.html.

Fig. 3.4   CONNECT2SEA website. Produced by the CONNECT2SEA project (2013–2016) (http:// www.connect2sea.eu/home.html)

control over their data and trust relations. The work also results in more resilient critical infrastructures and services. The project intends to reach out beyond its constituency to the broader public and to policy makers [11].

3.2.3.2   European (Horizon 2020) Projects in the Area of Cybersecurity

This section lists the most important recent European projects in the area of cybersecurity and privacy. This list is not intended to be exhaustive, but it aims at undertaking a review by providing the most reliable examples of Horizon 2020 projects:

- **Secure Hardware-Software Architectures for Robust Computing Systems (SHARCS)**
  SHARCS proposes a proactive response to cybersecurity instead of tackling cyberattacks afterwards. There have been an increasing number of cyberthreats in the past few years that need cooperation and anticipation. SHARCS intends to constitute a platform that will be used to design, implement and demonstrate secure-by-design applications and services by analyzing the layers of hardware & software

of a computing system and quantify the performance in real world applications. The application covers diverse domains including medical, cloud and automotive.

- **UNFRAUD**
  UNFRAUD is an advanced, online, anti-fraud, reliable software platform[13] that detects fraudulent methods via deep learning neural networks.

- **CANVAS**
  Canvas consortium intends to build a community that will unify the distributed efforts by technology developers, legal/philosophical scholars and empirical researchers in the field of cybersecurity. The main goals of CANVAS focus on structuring the current knowledge coming from these different perspectives, designing a network of information dissemination across these different domains.

- **DISCOVERY**
  DISCOVERY intends to promote the dialogue and cooperation between Europe and North America (US and Canada) with regard to funding mechanisms, ICT policy and cybersecurity.

- **AEGIS**
  AEGIS (Accelerating EU-US Dialogue for Research and Innovation in Cybersecurity and Privacy) aims to strengthen international dialogues on cybersecurity and privacy in order to facilitate the exchange of viewpoints, policies and best practices to accelerate EU-US cooperation in cybersecurity and privacy research and innovation.

- **SAINT**
  SAINT plans to fight the cybercrime by analyzing the ecosystem of cybercriminal activity and revealing evidence on privacy issues and Deep Web practices. Additionally, SAINT plans to build a framework in order to advance the measurement techniques of metrics of cybercrime by adding innovative models and algorithms.

- **REDSENTRY**
  redSentry plans to build a real-time networking monitoring framework that will detect and manage any type of cyber threat faced by the financial services sector. The network monitoring of cyber threats will allow the financial sector to manage their responses automatically and proactively.

- **Cybersecurity Behavioral Toolkit (CLTRe)**
  The Cybersecurity Behavioral Toolkit—CLTRe intends to measure and correct the security level of an organization by checking the behavior of its employees. CLTRe will store these actions and will generate a trend analysis that will give insight into the behavioral change over time.

- **3ants**
  3ants intends to build a cybersecurity framework against piracy and digital distribution of counterfeited goods by utilizing machine learning techniques. The framework is following crawler pirated contents in the Internet and forces the withdrawal of the goods by applying geolocation techniques.

- **CS-AWARE**
  CS-AWARE plans to tackle the issue of cybersecurity in commercial companies,

---

[13]http://www.unfraud.com/en.

NGOs, and governmental institutions regarding information sharing. They suggest a cybersecurity awareness solution for public administrations by applying analysis of the context. This analysis will give automatic incident detection and visualization and will promote exchange of information with EU level NIS authorities (CERTs).

- **CYBECO**
  CYBECO proposes to develop a framework for cyberinsurance, by combining multidisciplinary methods from Behavioural Economics, Statistics, Game and Decision Theory and build new models that will be coupled within a prototype software architecture (CYBECO Toolbox 2.0).

- **STORM**
  Storm is a Cyber Risk Management system which serves evidence based on metrics that can describe the cyber risk in real time, regarding business asset level. They leverage an efficient risk model, that enhances user experience and replies to queries like:

  - How much cyber risk do we have?
  - What should my customer pay for cyber insurance?

  Storm promotes collaboration on cyber risk management, by information sharing, advancement of analysis software, and interaction with current user involvement.

- **cyberwatching.eu**
  cyberwatching.eu intends to instrument and maintain an EU Observatory that will watch R&I initiatives on cybersecurity and privacy. Their main outputs include: the implementation of an online catalogue of services regarding cybersecurity and privacy, a cluster tool, webinars, annual workshops, concertation meetings, regional SME workshops, cluster reports, white papers, roadmaps and sustainability through cybersecurity and privacy marketplace.

- **Protection Beyond Operating System(ProBOS)**
  Protection Beyond Operating System(ProBOS) proposes a next generation cybersecurity protection system, by applying Artificial Intelligence approaches, that will offer a high level of protection to governments and private organizations.

- **Traditional Organised Crime and the Internet: The changing organization of illegal gambling networks (OCGN)**
  OCGN plans to examine online gambling and extortion networks and organized crime that is attracted around these. They aim to develop and validate tools and methods to simplify the management of cyberthreats in online gambling; define standards of information management and dissemination of information to cybersecurity centres from online gambling sector in EU.

- **reliAble euRopean Identity EcoSystem (ARIES)**
  reliAble euRopean Identity EcoSystem (ARIES) intends to build a schema for reliable secure and privacy-respecting physical and virtual identity management by utilizing new technologies offering highest level of quality in eID. Their goal is to use virtual and mobile IDs cryptographically derived from eID documents that will not allow any identity theft by utilizing biometric features as well.

- **SISSDEN**
  Based on the experience of Shadowserver, known in the security community for achievements in the fight against botnets and malware and close cooperation with Law Enforcement Agencies, network owners and national CSIRTs, SISSDEN aims to improve the cybersecurity situational awareness of EU entities through building a large network of passive sensors, providing free-of-charge victim notification feeds, and developing innovative data collection and analysis tools, with a strong focus on effective botnet tracking.

- **CyberWiz**
  CyberWiz is determined to build a tool in order to understand cybersecurity levels across a large complex number of interconnected systems in ICT environments of critical infrastructure (energy distribution). The main goals focus on the proactive management of cybersecurity, the generation of a vulnerability "heat-map" for the configuration of the involved parts, along with a user friendly visual contrasting of diverse alternatives and a validation of the tool in two pilots (energy utilities in Sweden and Germany).

- **European Network for Cybersecurity (NeCS)**
  NeCS was structured to deal with the increasing needs for new, highly qualified researchers in the area of cybersecurity who are called to be challenged with the evolvable cyberthreats and information security vulnerabilities. The project will focus on a multi-sectorial/disciplinary approach including participants both from academic and non-academic sector.

- **Wide-Impact cyber SEcurity Risk framework (WISER)**
  WISER (Wide-Impact cyber SEcurity Risk) framework aims to build a cyber-risk management real-time framework that will estimate, oversee and mitigate cyber risks by implementing nine experiments (Early Assessment Pilots: EAPs). These experiments will be the basis to build an advanced risk management system that will be tested & validated in 3 Full Scale Pilots (FSPs).

- **LocationWise**
  LocationWise is a cloud based location verification system that intends to lower fraud costs acquired by banks. The idea is to get the user's location (through his mobile) and reduce the number of fraudulent transactions and false positives.

- **MEDIA4SEC**
  MEDIA4SEC's main intention focuses on the exploration of the challenges and ethical considerations social media use for public security. This will be done by communication and dissemination activities to make the public aware of all the cyber risks in social media like trolling, cyberbullying, threats, or live video-sharing of tactical security operations. Additionally, they plan to explore the usage of social media for public security purposes highlighting ethical, legal and data-protection-related issues.

- **Proactive Risk Management through Improved Cyber Situational Awareness (PROTECTIVE)**
  PROTECTIVE is designed to improve an organisation's ongoing awareness of the risk posed to its business by cybersecurity attacks. PROTECTIVE aims at the defense mechanisms of an organization towards incoming cyber attacks, malware

outbrakes, and other security issues through two main actions: the development of the computer security incident team's (CSIRT) threat alertness via security monitoring; and the ranking of critical alerts depending on the potential disturbance that can be caused. PROTECTIVE intends to give solutions to public domain CSIRTs and SMEs. For the evaluation process they will conduct two pilots.

- **Collaborative and Confidential Information Sharing and Analysis for Cyber Protection (C3ISP)**
  C3ISP intends to create a cybersecurity, analysis and protection, management framework that will enable the collaborative and confidential sharing of information. The C3ISP will improve the detection of cyber threats while maintaining the confidentiality of information within a collaborative, multi-domain environment. They indend to enable the transitions from high level descriptions (close to natural language) to data usage policies which can be enforced in the system; in order to be able to work directly on encrypted data, they deal with the most applicable data protection schemes.

3.2.3.3 Pilot Projects

Four pilot projects achieved financing at the 2018 Horizon 2020 cybersecurity call SU-ICT-03-2018—"establishing and operating a pilot for a European Cybersecurity Competence Network and developing a common European Cybersecurity Research & Innovation Roadmap". These projects are expected to strengthen the EU cybersecurity capacity and overcome future cybersecurity challenges. The projects are expected to cooperate closely and coordinate their activities in order to advance the performance of cybersecurity research, innovation and deployment. These pilot projects bring together more than 160 partners from 26 EU Member States. The overall EU investment in these projects will be more than EUR 63.5 million. These projects are briefly described below.

CONCORDIA

CONCORDIA is a four-year research and innovation project, which aims to increase the effectiveness of EU security. The project, launched in January 2019, is coordinated by the CODE Research Institute from the Bundeswehr University Munich and involves 46 partners. CONCORDIA aims to strengthen European security capabilities and to provide a digital society and economy. The project is expected to develop innovative, market-based solutions to protect Europe from cyberattacks and establish an European Education Ecosystem for Cybersecurity. The project should be an instrument to promote research, market innovation, skills building, and a research roadmap for cybersecurity in Europe.[14]

ECHO

The ECHO project (the European network of Cybersecurity centres and competence Hub for innovation and Operations) is a four-year project that comprises 30 partners

---

[14]www.concordia-h2020.eu.

from 15 EU Member States and Ukraine and is coordinated by the Royal Military Academy of Belgium. The project is expected to deliver an organized and coordinated approach to strengthen pro-active cyber defense in the European Union through effective and efficient cross-sector collaboration. ECHO partners are expected to develop, model and demonstrate a network of research and competence in the field of cybersecurity, with a centre of research and competence at the hub.

The main goal of the project is to organize and optimize the currently fragmented cybersecurity activities across the EU. The Central Competence Hub will serve as the focal point for the ECHO Multi-sector Assessment Framework enabling analysis and management of multi-sector cybersecurity dependencies, including:

- Development of cybersecurity technology roadmaps;
- Creation of an ECHO Cybersecurity Certification Scheme aligned with ongoing EU efforts;
- Provision of an ECHO Early Warning System;
- Operation of an ECHO Federation of Cyber Ranges;
- Delivery of the ECHO Cyberskills Framework, with training materials and training events;
- Management of an expanding collection of Partner Engagements.[15]

SPARTA

The SPARTA project (Strategic Programs for Advanced Research and Technology in Europe) is led by CEA, and assembles a balanced set of 44 actors from 14 EU Member States. SPARTA is a cybersecurity competence network, with the objective to collaboratively develop and implement research and innovation actions. The four initial research and innovation programs are the following:

- **T-SHARK—Full-spectrum cybersecurity awareness**—to increase the range of threat understanding, from the current investigation-level definition, up to strategic considerations, and down to real-time events
- **CAPE—Continuous assessment in polymorphous environments**—to improve assessment processes to be able to perform continuously in the HW/SW lifecycles, and under changing environments
- **HAII-T—High-Assurance Intelligent Infrastructure Toolkit**—to manage the heterogeneity of the IoT by providing a secure-by-design infrastructure that can offer end-to-end security guarantees
- **SAFAIR—Secure and fair AI systems**—to evaluate security of AI systems, producing approaches to make systems using AI more robust to attackers' manipulation. Furthermore, the goal is to make AI systems more reliable and resilient through enhanced explainability and better understanding of threats.[16]

---

[15]www.echonetwork.eu.

[16]www.sparta.eu.

CyberSec4Europe

CyberSec4Europe (Cyber Security for Europe) is a research-based project involving 43 participants from 22 EU Member States and Associated Countries. As a pilot for a Cybersecurity Competence Network, it will test and demonstrate potential governance structures of a network of competence centres using examples of best practices from the expertise and experience of participants. The project has identified key demonstration cases in different industrial domains, including finance, healthcare, transportation and smart cities. These address prominent research areas in both the public and private sectors that correspond closely to real-world issues, cyber threats and security problems. Another important outcome of the project is expected to be the development of a framework model of cybersecurity skills that will be used as a reference point by providers of educational services, employers and citizens.[17]

3.2.4 Major Collaboration Centres

The section very briefly describes the main research coordination centers and initiatives in the area of cybersecurity and privacy, such as European cybersecurity competence centre and network and Digital Innovation Hubs. The section is not intended to describe individual institutions involved in research in the area of cybersecurity.

3.2.4.1 Digital Innovation Hubs

Digital Innovation Hub (DIH) is a concept of the European Commission, which has been proposed in several phases. First, Competence Centres had been expected to develop the concrete technologies to foster digital transformation. Next, on the basis of various analyses, the concept of Digital Innovation Hubs has arisen. Now, DIHs are aimed at building the ecosystem of digital innovation based on connecting various environments and sectors and knowledge, experience and technology transfer. The EU supports the collaboration of DIHs to create an EU-wide network in which companies can access competences and facilities not available in the DIH in their region. This network will lead to knowledge transfer between regions and will be the basis for economies of scale and investments in hubs. DIHs ought to be:

- one-stop-shop: one-stop-shops where companies—especially SMEs, startups and mid-caps—can get access to technology-testing, financing advice, market intelligence and networking opportunities;
- a competence centre, which acts as an intermediary between companies and investors and provide knowledge, expertise and technology to companies in the region;

---

[17] www.cybersec4europe.eu.

- a contact point in a state, which strengthens innovation and creates an area to collaborate for various stakeholders.

To help DIHs to effectively collaborate and network, the European Commission set up the European catalogue of DIHs.[18]

3.2.4.2 European Cybersecurity Competence Network and Centre

In September 2018, the European Commission proposed the creation of a Network of Cybersecurity Competence Centres and a new European Cybersecurity Industrial, Technology and Research Competence Centre. The aim of the proposition was to help the EU to maintain and develop technological and industrial capabilities related to cybersecurity necessary to secure cyberspace and increase the competitiveness of the EU cyber industry and transform cybersecurity into a competitive advantage of other European industry sectors. The pilot projects, described earlier in Sect. 3.2.3.3 are meant to implement these aims. The proposal has created a Network of National Coordination Centres, a Cybersecurity Competence Community and a European Cybersecurity Industrial, Technology and Research Competence Centre [5]. The Competence Centre will facilitate and help to coordinate the work of the Network and nurture the Cybersecurity Competence Community, driving the cybersecurity technological agenda and facilitating common access to the expertise of national centres. The tasks of the Competence Centre are the following:

- facilitate and assist in coordinating the work of the National Coordination Centres Network and the Cybersecurity Competence Community;
- increase cybersecurity capabilities, knowledge and infrastructures in the service of industries, the public sector and research communities;
- contribute to the wide deployment of state-of-the-art cybersecurity products and solutions across the economy;
- improve understanding of cybersecurity and contribute to reducing skills gaps in the Union related to cybersecurity;
- contribute to the reinforcement of cybersecurity research and development in the Union;
- enhance cooperation between the civil and defence spheres with regards to dual use technologies and applications in cybersecurity;
- enhance the synergy between the civil and defence dimensions of cybersecurity in relation to the European Defence Fund.

Each Member State will designate one National Coordination Centre. They will function as a contact point at the national level for the Competence Community and the Competence Centre. The National Coordination Centres shall have the following tasks:

- supporting the Competence Centre in achieving its objectives and in particular in coordinating the Cybersecurity Competence Community;

---

[18]http://s3platform.jrc.ec.europa.eu/digital-innovation-hubs-tool.

- facilitating the participation of industry and other actors at the Member State level in cross-border projects;
- contributing, together with the Competence Centre, to identifying and addressing sector-specific cybersecurity industrial challenges;
- acting as contact point at the national level for the Cybersecurity Competence Community and the Competence Centre;
- seeking to establish synergies with relevant activities at the national and regional level;
- implementing specific actions for which grants have been awarded by the Competence Centre, including through provision of financial support to third parties;
- promoting and disseminating the relevant outcomes of the work by the Network, the Cybersecurity Competence Community and the Competence Centre at national or regional level;
- assessing requests by entities established in the same Member State as the Coordination Centre for becoming part of the Cybersecurity Competence Community.

3.2.5 Other Activities

This section provides examples of the "operational" (to some extent) activities, other than research and innovation projects. This section covers cybersecurity and privacy activities (on the basis of a few examples from various countries), which do not necessarily result from the national strategy, but are in some sense relevant.

In Greece, the Ministry of Digital Policy, Telecommunications and Media[19] include in its missions the development and implementation of national policy, the contribution in the development of an appropriate institutional framework at European and international level, towards the development of high quality communication and digital politics. The ministry targets at the promotion of security in telecommunication, digital information and information media as well as the development of telecommunications' market guidelines.

In Poland, the Research and Academic Computer Network—National Research Institute (NASK-PIB) has published a web portal, CyberPolicy.[20] The CyberPolicy portal aims at providing knowledge about cybersecurity from the strategic, regulatory, organizational and practical perspectives. The main mission of the portal is to gather and disseminate information regarding new regulations, programmes, strategies, good practices and initiatives directly connected with the cybersecurity's ecosystem. However the portal is currently not very comprehensive and only available in Polish, but the dynamic growth of the portal is very promising.

In Belgium, The Cyber Security Coalition is a unique partnership between players from the academic world, the public authorities and the private sector to join forces

---

[19]http://mindigital.gr/.

[20]https://cyberpolicy.nask.pl/.

in the fight against cybercrime. Currently more than 50 key players from across these 3 sectors are active members contributing to the Coalition's mission and objectives.

In France, the OSSIR association (Information System and Network Security Observatory),[21] founded in 1996, gathers people interested in cybersecurity and fosters exchanges and scientific presentations through four regional working groups. Each year, it also organizes the JSSI (Information Systems Security Day) with conferences on a specific topic of cybersecurity. Another association, the CLUSIF (French Information Security Club),[22] gathers professionals from the information security area and aims at fostering exchanges of ideas through working groups, thematic conferences and publications, freely available on its website. As of today, the CLUSIF membership is comprised of more than 300 companies covering 15 different lines of business. Regarding cybersecurity evaluation, it is being delegated to CESTIs, which are independent IT Security Evaluation Centers. A CESTI must be authorized by the French national cybersecurity agency (ANSSI) in order to deliver certifications for specific domains, such as conformance to the Common Criteria model. Currently, a dozen companies in France have been authorized as CESTIs. More recently, the French national scientific research center (Centre National de la Recherche Scientifique—CNRS) has created a network of cybersecurity researchers. The *Groupe De Recherche* (GDR) *Sécurité Informatique*[23] animates the community of French researchers in the domain of cybersecurity with an exhaustive coverage in terms of topics. It organizes national days in June every year, and provides a label to national conferences, and international conferences organized in France. It is the main communication tool for academic researchers interested in cybersecurity.

3.3 The Strong and Weak Points

This section is the first part of a summary—it provides a general overview of strong and weak points for both regions. In this section, we provide conclusions drawn on the basis of available information. This section and the next section have been prepared on the basis of a number of sources of information. First of all, the outcomes of questionnaires collected during and after the first EUNITY workshop in Japan and during the second workshop in the EU have been deeply analysed. While the number of questionnaires collected is not very large (several dozens), their significance is very high, because of the fact that they were provided by high-level experts in the area of cybersecurity. In addition, every expert represents not only one's own point of view, but also the point of view of her institution. Secondly, the observations and experience of all partners of the consortium have been used to broaden the landscape of the problems indicated in questionnaires and highlight the common ones. Another very important source of information is the review of the strategies, projects and

---

[21] https://www.ossir.org/.

[22] https://clusif.fr.

[23] https://gdr-securite.irisa.fr/.

programs with regard to research and innovation in the area of cybersecurity and privacy, as well as relevant financing mechanisms, which have been provided in the previous sections of this chapter.

It is worth noting that in many aspects of research in cybersecurity and privacy, the strong and weak points are common for both regions. In case of significant differences, the most important distinctive features are provided.

This section focuses more on the organizational aspects of research in cybersecurity and privacy. The second (and the last) part of the summary (which is provided in Sect. 3.4) focuses on more technical aspects of the area of cybersecurity and privacy, and states the main similarities and differences of both regions.

3.3.1   Strengths

3.3.1.1   Establishment of the Cybersecurity Strategy

During the review of the cybersecurity strategies, we analysed strategic documents collected at various levels (EU, national, etc.). Some of them point out the most important technical areas of research. In general, the strategic vision of cybersecurity is an advantage both in the EU and Japan.

The table lists common areas in the field of research and innovation from the perspective of strategic documents.

For more detailed comments, refer to the earlier tables: Table 3.2 (EU) and Table 3.4 (JP) (summarized in Table 3.5).

In the strategic perspective, there is a high level of awareness on both sides. Small improvements are only possible at Member States' level for the EU side.

Responses to the questionnaires were provided by many of the representatives of industry, government, research & academia and CSIRT community. The level of knowledge about national / international / sectorial research strategies, roadmaps and strategic agendas focused on cybersecurity or privacy is, in general, lower in case of representatives of the industry. As a consequence, it makes it difficult, for example, to encourage the industry to aim towards strategically relevant directions. On the contrary, many representatives state that such strategic directions pointed out by these strategic documents are well aligned with the needs of the industry. The most interesting opinions are the following:

- There is a need to have reliable metrics in cyber hygiene, for the common base of discussion and mitigation. Therefore, strategic documents are very important.
- Although the use of IT has already progressed widely in various fields, the regulation and the government policy are still able to have a great effect on the activities of the private sector and the citizens. Because of that, promoting public-private cooperation in the use of IT, cybersecurity and privacy protection will be critical for wholesome economic growth.

Table 3.5 The comparison of fulfilled objectives in cybersecurity strategies between the EU and Japan

Objective	The EU		Japan
	The EU level	The Member States' level	
Set the vision, scope, objectives and priorities	✓	–/ ✓	✓
Follow a national risk assessment approach	✓	–/ ✓	✓
Take stock of existing policies, regulations and capabilities	✓	–/ ✓	✓
Develop a clear governance structure	✓	✓ / –	✓
Identify and engage stakeholders	✓	✓ / –	✓
Establish trusted information-sharing mechanisms	✓	–/ ✓	✓
Develop national cyber contingency plans	✓	✓	✓
Protect critical information infrastructure	✓	✓	✓
Organise cyber security exercises	✓	✓	✓
Establish baseline security requirements (measures)	✓	✓	✓
Establish incident reporting mechanisms	✓	✓	✓
Raise user awareness	✓	✓	✓
Strengthen training and educational programmes	✓	✓	✓
Establish an incident response capability	✓	✓	✓
Address cyber crime	✓	✓	✓
Engage in international cooperation	✓	✓	✓
Establish a public/private partnership	✓	✓/–	–
Balance security with privacy and data protection	✓	✓/–	✓
Institutionalise cooperation between public agencies	✓	✓	✓
Foster R&D	✓	✓	✓
Provide incentives for the private sector to invest in security measures	✓	✓ / –	✓

- Where there is absolutely no need, services cannot be sold on the market. However, pursuing only needs, will not yield more advanced business markets, making strategic documents even more relevant.
- Lately, research strategies require programs to involve companies interested in technologies from the point of view of exploitation. Thanks to those companies, the industry needs become input to research programs.

3.3.1.2 Declared Focus on Cybersecurity and Privacy

Both EU and Japan are fully aware of the current importance of cybersecurity. Both regions emphasize the need to strengthen related areas, primarily in research, development and innovation contexts. At the level of the EU as a whole, as well as in Japan, a strong focus on research in the area of cybersecurity can be noticed in strategic documents. This is also evidenced by the growing number of funds dedicated to conduct research in these areas. But, due to heterogeneity in the EU at the Member States' level, small differences could be perceived. First of all, not all EU countries put exactly the same emphasis on the area of cybersecurity. More importantly, the way of financing research in this area can be also different at the national level. In general, however, many efforts have been made to encourage strategic investments in the area to foster progress, hoping to catch up with countries more advanced in this area (both in the case of research and industry in cybersecurity).

3.3.2 Weaknesses

3.3.2.1 Opposition Between Industry and Research

This subsection discusses the obstacles in transferring outcomes of research into technologies and products as well as oppositions between industry and research. The analysis is based on the outcomes of questionnaires as well as on our own observations. The most interesting finding is that the main problems are common for both regions (the EU and Japan).

A very important problem is the ability and time needed to transfer the outcomes of research into technologies and products. This problem is not specific to cybersecurity, but in this area it can be especially challenging. The main indicated obstacles are outlined and briefly explained below:

- **Different focus**. Research and innovation usually focus on cutting-edge solutions, but the industry focuses on well-understood and tested (also, often well supported) and, in many cases, widely-adopted solutions. In other words, the industrial side is becoming conservative, because there is little consciousness to adopt new solutions progressively.

- **Different needs**. Even if research was able to provide a solution, which would be ideal in some ways, it could be still impractical. For example, a solution, which would be able to detect all network attacks, could be impractical if it was bounded to high level of false positives.
- **Different aims**. Sometimes it is really difficult to draw researcher's attention to the hidden needs of existing solutions.
- **Different assumptions**. It is difficult to use outcomes of research in the real world, because outcomes of research are often effective only under certain circumstances, which are not always met in the real world.
- **Law limitations**. New solutions are often not legally viable because of the fact that the existing regulation establishes limits to certain practices that would be necessary for such new technologies to be deployed and used. For example, the data protection regulation represents a barrier to numerous data-processing based business models. Even cybersecurity solutions are affected by data protection regulations (IP addresses, normally managed in volumes for cyber threat analysis, are considered "personal data" and therefore they cannot be collected or stored as freely as it would be necessary). When it comes to research activities, GDPR allows the use of less stringent data protection rules for scientific data processing. However, the transformation of research outcomes into products may face the challenge of stricter privacy rules within the commercialization phases.
- **Fragmented regulations across the different markets**. Due to the fact that every solution has to be implemented in a certain environment, it is not strictly applicable to find a solution which will satisfy the needs of all potential adopters. For example, due to the various legal frameworks in different countries, there is a possibility that a solution, which is objectively very good (and proven by research outcomes) cannot be adopted due to legal prohibition. It is worth noting that a fragmented European landscape with regard to cybersecurity and privacy somehow limits research exploitation, but this situation is changing due to efforts to harmonize regulations. For instance, in security-related projects involving law enforcement, the applicable legislation often differs between the research project phase and the end-use or exploitation, which generates loopholes. The disparate regulatory frameworks across Europe create much inefficiency and produce a fragmented market that makes it hard and costly for new technologies to thrive as viable businesses. A very important difference occurs between Japan and European markets. For example, in Japan, scanning tools like SHODAN or Shadowserver are in the grey zone of regulation, and because of that it is almost impossible for the industry to use such tools. Instead, the government of Japan gave such authority to NICT through the ministerial order from the MIC, enabling the agency to scan IoT devices in Japan for vulnerabilities and notify its owners in collaboration with ISPs.
- **Conflict between theoretic model and reality**. Outcomes of research are in many cases based on a theoretical analysis that can actually lead to a conflict between reality and theory, due to the limitation of the model used.
- **Limited resources**. Even the best solution cannot be implemented in all situations because of various limits, for example due to the cost of such solution and lack

of economic justification. On the other hand, in practice (in the industry) other parameters than only performance are also taken into account, for example user-friendly UI or ease-of-use.

- **Different limitations**. Outcomes of research can indicate that a solution should be deprecated, but in practice, due to various reasons, the solutions cannot be easily replaced by another solution (for instance, another solution does not exist or such replacement would be too costly). For example, even though outcomes of research recommend or even require immediate implementation of more secure encryption algorithms, it is not adoptable because of the limitations of the existing system.
- **Time of adoption**. Every company has a business model. It is often difficult to incorporate new technologies into the business model smoothly. For example, even if a good new cyber attack detection technology has been developed, it is very difficult for existing monitoring companies to properly respond to detected alerts due to the lack of security knowledge. It usually requires hiring additional people understanding security, which is not taken into account in the current business model.

Interestingly, such obstacles are widely indicated by many of the surveyed experts. Some of these obstacles are also confirmed by the SRIA document [9].

We can indicate the following propositions to reduce these obstacles:

- industry involvement in research,
- practical experience of the researchers,
- good definition of mutual expectations between research and industry,
- law harmonization, which can bring the effect of synergy, and because of that, research outcomes could be adopted and implemented more easily.

All of the above aspects are very important, but in the case of cybersecurity, industry involvement in research and good practical field-knowledge of the researchers are even more important than in other areas of industry.

Problems are in general very similar between the EU and Japan. One typical issue in both EU and Japan is the reliance of governments on foreign-made security appliances, in many cases exclusively.

3.3.2.2   High-Level Cybersecurity's Personnel Shortage

During the analysis, the lack of high-qualified personnel in the area of cybersecurity was highlighted. The analysis provided is based on the outcomes of questionnaires, the strategic documents regarding cybersecurity, as well as on our own observations. The most interesting fact is that both regions suffer from common deficiencies in terms of workforce (the EU and Japan).

On the basis of the received responses to the questionnaires, we can state that education remains very important at every possible level. For example, one expert states that there is often a need to perform courses from the bottom up, because sometimes even IT professionals lack basic knowledge essential to cybersecurity (for example, the basics of the UNIX command line and other related fundamentals). There is also an indicated need for education in handling information within an organization. On the other hand, many responders from both regions state that there is a very high need for education of average citizens to promote and improve awareness of cybersecurity. One of the responders also states that there is a need for more education on top of the research activities undertaken by academia and specialized centers. This is indeed particularly important as research and innovation should not be separated from education and professional training.

As a matter of fact, Working Group 5 (WG5) at the European Cyber Security Organisation (ECSO) is devoted to aspects of education in the area of cybersecurity. Specifically the working group focuses on the following issues[24]:

- *Increase education and skills on cybersecurity products and safe use of IT tools in Member States for individual citizens and professionals.*
- *Develop cybersecurity training and exercise ecosystem leveraging upon cyber range environments*
- *Support awareness-raising and basic hygiene skills*

As the *WG5 Position Paper* states: *there is clearly a lack of qualified cyber security professionals and a need for a rise in the overall level of cyber awareness. The demand for cyber specialists and experts is greater than the supply and this is making society and organisations increasingly vulnerable. Governments, associated with different stakeholders, should tackle the cyber security skills gap through more education and training accredited offers. This has become a very high priority for the European Commission, which is addressing this skills gap through several programmes, including: A New Skills Agenda for Europe: Working together to strengthen human capital, employability and competitiveness* [12].

Moreover, the document [12] also states that: *with regards to cyber security, we are already in a crisis for not producing enough skilled experts that the industry is desperately looking for and stakeholders also lack the knowledge to understand the nature of the cyber domain.*

Very similar problems have been noticed on Japan's side, as the *Cybersecurity Strategy* of Japan [7] states *development and assurance of cybersecurity workforce* as one of the aims, because of the fact that *cybersecurity workforce development is a pressing task for Japan, as there is a critical domestic shortage of cybersecurity experts, both in quality and quantity.*

Of course, the shortage of professionals affects all fields of cybersecurity, but it could be especially devastating in the field of research and innovation.

3.3.2.3 Lack of Coordination of Research Actions on Various Levels

During the analysis of available materials (such as information about projects and programs, as well as financing mechanisms), including the review of questionnaires

---

[24]https://www.ecs-org.eu/working-groups/wg5-education-awareness-training-cyber-ranges.

and on the basis of our own observations, the improper coordination of research actions in the area of cybersecurity at various levels (international, national, sectorial, etc.) has been noticed. Inevitably, the lack of coordination is more evident in the EU landscape (due to the larger market but not due to the lack of actions which aim at synchronization). In the EU cybersecurity landscape, research is conducted in a wide range of programs and is funded from different sources, as it has been indicated in Sect. 3.1. Unfortunately, there is a possibility that similar actions and projects can be pursued by various entities at the same time, which may lead to inefficiencies and wasted opportunities to achieve synergy. Contrary to other issues mentioned in this section, which are common to both regions, this problem is significantly more severe on the EU side.

3.3.2.4 Lack of Strong Global Cybersecurity Enterprises and Solutions Originating in the EU and Japan

During the analysis of questionnaires as well as on the basis of own observations, we have noticed that one of the main problems in the cybersecurity market is the insufficient presence of strong European or Japanese enterprises which could provide globally competitive products. Because of that, the EU and Japan have to import such solutions from abroad. Of course, this is a problem in the context of industries, but it definitely has a strong impact on the research and innovation aspect as well (without high quality research with industrial or practical emphasis there is no chance to provide such solutions).

3.4 Common Interests Between the EU and Japan

In this section, we briefly describe the common interests between the EU and Japan which may create opportunities. During and after the workshops, the *EUNITY* team has collected answers to many of the problems regarding research and innovation aspects of cybersecurity.

3.4.1 Main Strategic Directions in Institutions

During the workshops, the *EUNITY* consortium attempted to obtain information from attendees about their institutions' most relevant strategic cybersecurity directions. The respondents were quite prolific and cited a wealth of research topics including (research topics are grouped by the authors of the book):

- **cyber threat intelligence**
  - high performance data analytics for cybersecurity,
  - operational security including tools for CSIRTs,

- information sharing,
- cyber attack visualization,
- threat analysis,

- **education, awereness and cyber range**

  - cybersecurity education and training,
  - security awareness training,
  - security testbed (cyber range),

- **data processing and privacy**

  - privacy and identity,
  - Big Data,

- **methods to enhance cybersecurity**

  - Artificial Intelligence for cybersecurity,
  - High Performance Computing for cybersecurity,

- **security services**

  - authentication/authorization,
  - digital certificate-based authentication infrastructure,

- **network security**

  - routing security,
  - file sharing methods,

- **cybersecurity in various domains**

  - IoT and cybersecurity in IoT,
  - Cloud Computing and cybersecurity in the cloud,
  - cybersecurity in critical infrastructures,
  - legal/policy on IT, IP, privacy, cybersecurity and cybercrime,
  - hardware security,
  - cloud computing,
  - social networks,

- **other**

  - cybersecurity technologies usable for the 2020 Tokyo Olympics.

In general, main topics listed by representatives of both regions are quite similar. It is worth noting that most of the listed topics are in line with general strategic cybersecurity objectives indicated in both regions. This demonstrates that the process of identification of needs performed during the development of the strategies was fruitful and on point. One interesting aspect is the fact that a few responders from Japan pointed out the topic of cybersecurity usability related to the 2020 Tokyo Olympic games.

3.4.2   R&I Cybersecurity Priorities and Current Directions

Based on the review of the strategies and research programs, as well as on the experts' opinion (e.g., collected using questionnaires) and our own observations, the most important cybersecurity research topics and main current research directions in the field are the following:

- **risk management and critical infrastructure protection**

  - risk measurement,
  - risk management on national and international levels,
  - risk evaluation (on the basis of information on incidents, such as malware infection, information leakage and so on),
  - cybersecurity measures in critical infrastructure fields such as electric power and railway,

- **cybersecurity in various technologies**

  - Internet of Things,
  - Smart Cities,
  - Industry 4.0,
  - Cloud Computing,
  - Big Data.

- **threat detection and threat intelligence**

  - new methods of detection of cyberattacks,
  - machine learning based threat detection,

- **cryptology design, techniques and protocols**

  - encryption,
  - bitcoin and digital currencies,
  - blockchain-based identity management services,

- **network security**

  - routing security,
  - IP security,
  - security of radio networks,

- **hardware and systems security**

  - secure embedded systems,
  - secure products and services,

- **cybersecurity measures at the Tokyo Olympic Games in 2020**.

The main research directions are very similar in both regions. The only difference is the focus on the Tokyo Olympic Games in Japan.

3.4.3 Identification of Threats

The most important common threats in both regions, which have been identified in majority by experts during the workshops, are the following (the threats were grouped by the authors of the book):

- **malware**

  - advanced malware (including sophisticated IoT malware),
  - ransomware,

- **APT**
- **cyber terrorism**

  - cyber terrorism (including online radicalization),
  - state-sponsored hacking activities,
  - industrial sabotage and safety-related cyberattacks,

- **network threats**

  - botnets and DDoS,
  - BGP hijacking,
  - vulnerable radio communication systems,
  - bluetooth threats (BlueBorne, BT-enabled critical systems),

- **cyber attacks for critical infrastructure**,
- **quantum cryptanalysis**
- **specific threats against various technologies**

  - IoT vulnerabilities,
  - drones,
  - 3D printers,
  - automotive,

- **lack of integration/cooperation between CERTs**,
- **data theft**

  - leakage of private information,
  - critical information theft and manipulation for economic activities,

- **social engineering**

  - identity theft/fraud,
  - spearphishing,
  - fake news propagated by strong origination, which leads public to misdirection,

- **poor cyber literacy**

It is worth noting that most of the threats are common for the EU and Japan. The most important and common areas are attacks targeting IoT as well as social engineering attacks. Big cybersecurity incidents are becoming increasingly frequent due to the massive appearance of IoT devices, which often include poor security features and weak infrastructures. Social engineering attacks (namely, attempting to attack the system by abusing the vulnerabilities of legitimate human users) are one of the most efficient paths to attack a system. Identity theft and ransomware are threats that make use of such social engineering attacks.

The issue of cybercrime attribution generates imbalances and issues in addressing foreign cybersecurity policies. Due to the difficulties in identifying hacking actors, homogeneous and common responses are often unachievable. Both regions are facing changes in terrorist threats, from cell/group-based terrorism to individual foreign fighters and lone wolves. For the latter, the Internet is a fundamental tool for dissemination, propaganda, radicalization and anonymous funding.

Some responders from Japan identify GDPR as a threat. When it comes to the invasion of privacy, the EU companies are required to report the details of breaches. Although there is similar regulation in Japan, the penalties and fines are significantly different, as we explained in Sect. 2.3.2.

3.4.4 Examples of Current Collaborations

On the basis of the questionnaires, a very important fact is that a vast majority of institutions have foreign collaborations with the EU and with some third parties other than EU. It can be concluded that such collaborations could be strengthened. The highest number of examples of collaborations can be seen in the area of research and joint projects. Exchanging staff is also a popular way of achieving international collaborations. The examples of such collaborations include:

- JPCERT/CC having many bilateral MoUs with European national CSIRTs, and exchanging incident information when necessary,
- JPCERT/CC being a member of APCERT[25] since its founding, and having contacts with Asia-Pacific (especially South-Eastern Asian) countries/economies,
- JIPDEC cooperating with ETSI (European Telecommunications Standards Institute) in the field of Internet trust.

3.4.5 ICT Areas Which Need Collaboration Between EU and Japan

The ICT areas other than cybersecurity and privacy which need collaboration, as indicated by responders are the following:

- communication technologies including 5G,
- NGN (Next Generation Network),

---

[25]https://www.apcert.org.

- IoT (Internet of Things) and cyber-physical systems, including drones,
- NFV (Network Function Virtualization),
- Big Data,
- HPC (High Performance Computing),
- AI (Artificial Intelligence) and ML (Machine Learning),
- VR/AR (Virtual/Augmented Reality),
- e-Health,
- smart cities,
- robotics,
- computer science education,
- distributed OS,
- interoperability between eID (government-issued digital certificate for citizen) in the perspective of EU and Japan (in regulation, policy and technical aspects).

One of the respondents mentioned something quite interesting: any systemic question that knows no borders (e.g., climate change, air/sea pollution) to which ICT may be part of the solution. On the whole, the need for collaborations is a mirror of the identified threats.

3.4.6 Areas Which Need the Most Collaboration

In this subsection, all the areas of cybersecurity which need collaborations are indicated:

- **education and awareness**
  - education for various audiences,
  - enhancing security awareness,
  - development of human resources,
  - promoting the exchange of personnel,

- **standards and regulations**
  - harmonization on standards and regulations among government and industrial associations,
  - guidelines by industry sector,
  - sharing best practices regarding cybersecurity,

- **information sharing**
  - sharing environments to monitor attacks,
  - sharing security intelligence among security vendors/organizations,
  - continuous information feeds on web sites, e.g., blogs or whitepapers,
  - continuous exposure in conferences/exhibitions,
  - continuous workforce activities, e.g., industry ISAC.[26]

---

[26]ISAC stands for *Incident Information Sharing and Analysis Center*.

Other activities which could be performed together are as follows:

- maintain Interpol-like cooperation and non-aggression treaties,
- improve communication, information/data sharing, legal framework,
- harmonize legal and penal frameworks to ensure effective prosecution of cyber-criminals,
- reduce administrative procedures,
- intensify collaboration between CERT/CSIRT teams,
- promote joint initiatives (including meetings and workshops).

Certainly boosting the responsiveness of Europe as a whole and fostering cooperation and coordination in cybersecurity between Member States and Japan is a very important issue. There is a need of industry-government cooperation and global collaboration to exchange sensitive data and to enlarge the cooperation to as many countries and industry sectors as possible. Global collaboration shall not be only horizontal, i.e., limited to state entities, nations and international organizations. Rather, global cooperation should be both horizontal and vertical, i.e. also involving private entities and other stakeholders (academia for instance).

References

1. Communication from the Commission to the European Parliament, the Council, the European Economic and Social Committee and the Committee of the Regions A Digital Single Market Strategy for Europe. 6.5.2015. COM/2015/0192 final. https://eur-lex.europa.eu/legal-content/EN/TXT/?uri=celex%3A52015DC0192. Cited 1 Dec 2019
2. Joint Communication to the European Parliament, the Council, the European Economic and Social Committee and the Committee of the Regions Cybersecurity Strategy of the European Union: An Open, Safe and Secure Cyberspace. 7.2.2013. JOIN/2013/01 final. https://eur-lex.europa.eu/legal-content/EN/TXT/?uri=CELEX:52013JC0001. Cited 1 Dec 2019
3. Joint Communication to the European Parliament and the Council Resilience, Deterrence and Defence: Building strong cybersecurity for the EU. 13.9.2017. JOIN/2017/0450 final. https://eur-lex.europa.eu/legal-content/EN/TXT/?uri=CELEX:52017JC0450. Cited 1 Dec 2019
4. Communication from the Commission to the European Parliament, the Council, the European Economic and Social Committee and the Committee of the Regions Strengthening Europe's Cyber Resilience System and Fostering a Competitive and Innovative Cybersecurity Industry. 5.7.2016. COM/2016/0410 final. https://eur-lex.europa.eu/legal-content/EN/ALL/?uri=CELEX:52016DC0410. Cited 1 Dec 2019
5. European Commission: Proposal for a European Cybersecurity Competence Network and Centre. (2018). https://ec.europa.eu/digital-single-market/en/proposal-european-cybersecurity-competence-network-and-centre. Cited 1 Dec 2019
6. ECSO: POSITION PAPER Gaps in European Cyber Education and Professional Training. (2018). https://www.ecs-org.eu/documents/publications/5bf7e01bf3ed0.pdf. Cited 1 Dec 2019
7. The Government of Japan: Cybersecurity strategy. (2015). https://www.nisc.go.jp/eng/pdf/cs-strategy-en.pdf. Cited 1 Dec 2019
8. ECSO: Strategic Research and Innovation Agenda. (2017). https://www.ecs-org.eu/documents/publications/59e615c9dd8f1.pdf. Cited 1 Dec 2019

9. ECSO: European Cybersecurity Strategic Research and Innovation Agenda (SRIA) for a contractual Public-Private Partnership (cPPP). (2016). http://ecs-org.eu/documents/ecs-cppp-sria.pdf. Cited 1 Dec 2019
10. Akhgar, Babak and Brewster, Ben: Combatting Cybercrime and Cyberterrorism: Challenges, Trends and Priorities. (2016). DOI:https://doi.org/10.1007/978-3-319-38930-1
11. EUNITY Grant Agreement, Annex 1 Description of the action
12. European Commission: Digital Skills at the core of the new Skills Agenda for Europe. https://ec.europa.eu/digital-single-market/en/news/digital-skills-core-new-skills-agenda-europe. Cited 1 Dec 2019

Chapter 4
Industry and Standardization Aspects

The chapter focuses on the analysis of the output of clusters and industry associations. It also includes descriptions of national strategic efforts for creating or developing business strategies. With respect to standards, it covers a large scope of standard bodies.

4.1 Industry Activity Around Research

In this section, the different efforts to create business strategies within the cybersecurity industry will be evaluated by analyzing numerous associations and clusters that operate at a European, Japanese and Member State levels.

One of the bigger issues and risks of organizations in the last years is cybersecurity due to the growing threat of cyber-attacks. The more the technology grows the more business opportunities but also the more risks and threats. Furthermore, this interest has been supported by initiatives at political level in the European Union in order to ensure security and privacy in critical infrastructures and other sectors. Additionally, SMEs, which are the most prevalent type of companies in Europe and Japan, are one of the main targets of cyberattacks (and figures keep increasing) [1–3].

Even though the current situation is already critical and the expected future is worsening, still companies do not invest as much as they should in cybersecurity, both in technical solutions and training of their employees [5, 6] . This has created a common topic for companies that need to invest part of their benefits in cybersecurity research and solutions, resulting in the emergence of industrial organizations and initiatives that analyse and collect the key priorities of industry in terms of research and innovation with a global vision.

Furthermore, published studies from agencies such as ENISA (European Union Agency for Network and Information Security), among others, provide regular studies about new cybersecurity challenges that industry will need to face in the mid- to long-term [4].

© The Author(s), under exclusive license to Springer Nature Switzerland AG 2021   99
A. Felkner et al., *Cybersecurity Research Analysis Report for Europe and Japan*,
Studies in Big Data 75, https://doi.org/10.1007/978-3-030-62312-8_4

4.1.1 Methodology

This chapter is based on an analysis of the Strategic Research and Innovation Agendas (SRIA) of industrial associations and technology associations, as well as reports from ENISA, to understand the efforts and measures undertaken and challenges identified by the European Industry when it comes to cybersecurity. The analysed initiatives have been chosen considering several criteria: scope of action (cross-cutting technologies and vertical markets), impact and activity level.

Several of these organizations were established to manage different contractual public-private partnerships (cPPP) on different topics. A cPPP is an instrument co-funded by the European Union for research and innovation activities in crucial sectors of Europe's economy, bringing together companies, universities, research laboratories, SMEs and other organisations. These cPPPs develop strategic research and innovation agendas where the R&I priorities and challenges of the industry are structured and defined.

These partnerships are essential for the cybersecurity market since they promote and generate a real dialogue between industry and the European Commission, driving R&I priorities in the future and, in doing so, fostering market approach of the innovative technologies and, on the other hand, driving the needs of citizens and countries to the private sector. This way public and private needs and development can cooperate and find common grounds. As a result, these partnerships improve the European Cybersecurity Market competitiveness, making Europe a more attractive location for national and international companies to invest and innovate.

For the elaboration of this section, we have also analysed the studies conducted by ENISA in various sectors. ENISA is considered an important player that provides recommendations on cybersecurity for several industries, working together with operational teams for designing strategies that satisfy challenges identified in different areas. ENISA works horizontally in multiple sectors such as Cloud, Big Data, CSIRT, IoT, Data protection, Intelligent Infrastructures, Risk Management and Trusted Services.

4.1.2 Associations and Clusters at EU Level

There are several industry associations that have, among others, a strong focus on research activities. These are related with the ICT sector, and have ample capabilities to generate potential impact in the European Cybersecurity market, regardless of their main topic of work or industry. The Tables 4.1 and 4.2 summarize the associations we have considered for this study.

Table 4.1 European technology associations

Acronym	Name	Main industry
ECSO	European Cyber Security Organization	Cybersecurity service and product providers, operators of essential services (energy, telecom, finance)
BDVA	Big Data Value Association	Service operators, data management industry, technology providers, cybersecurity product providers
NESSI	Networked Software and Services Initiative	Software, services & data
EOS	European Organization for Security (EOS)	Cybersecurity industry, technology providers, solutions and services
ECSEL (incl. Artemis)	Electronic Components and Systems for European Leadership	Smart mobility, smart society, smart energy, smart health, smart production, design technology, cyber-physical systems, smart systems, safety and security

Table 4.2 European vertical sectors studied

Vertical sector	Industrial Associations and Organizations studied
Internet of Things (IoT)	Alliance for Internet of Things Innovation (AIOTI), ENISA reports
Communication infrastructures	5G Infrastructure Public Private Partnership (5G-PPP), European Telecommunication Network operators Association (ETNO)
Security	European Organization for Security (EOS)
Transport and mobility	ERTICO, European Road Transport Research Advisory Council (ERTRAC)
Manufacturing	European Factories of the Future Research Association (EFFRA)
Health	European Federation for Medical Informatics (EFMI), MedTech Europe Association

4.1.2.1 Technology Perspective Associations

European Cyber Security Organisation (ECSO)

ECSO is a non-profit organisation that represents the industry-led contractual counterpart to the European Commission for the implementation of the Cyber Security Public-Private Partnership. ECSO is composed of 219 organisations (large companies, public administrations, SMEs, regional clusters, etc.) from 28 countries. The main objective of this organization is to support all types of initiatives, that aim to promote European cybersecurity. This is done by:

- protecting the European Digital Single Market from cyber threats,
- cooperating towards the growth of a competitive cybersecurity and ICT industry,
- developing cybersecurity solutions for the critical steps of trusted supply chains.

ECSO defines 8 areas of interest for cybersecurity: ICT Infrastructure, Smart Grids, Transportation, Smart Buildings and Smart Cities, Industrial Control Systems, Public Administration and Open Government, Healthcare, Finance and Insurance.

The members of the organisation are involved in 6 working groups with different task objectives:

- WG1: Standardisation, certification, labelling and supply chain management. This working group addresses multiple issues such as labelling, standards for interoperability, increased digital autonomy, etc.
- WG2: Market deployment, investments and international collaboration. Aimed at supporting members to improve market knowledge, facilitating investment capabilities and promoting international trade.
- WG3: Sectoral demand. This group brings together cybersecurity stakeholders in order to engage directly with users, linking supply and demand.
- WG4: Support to SMEs, coordination with countries.
- WG5: Education, awareness, training, cyber ranges. This working group focuses on cybersecurity training.
- WG6: Strategic Research and Innovation Agenda. Coordination of cybersecurity results and activities of R&D projects.

It is worth noting that working group 6 published its last Strategic Research & Innovation Agenda (SRIA) on June 2017 [7] where their priority challenges around cybersecurity are presented:

- Remove trust barriers for data-driven applications and services: through data security and privacy technologies, distributed identity and trust management and user-centric security & privacy.
- Maintain a secure and trusted ICT infrastructure in the long-term: focusing on ICT infrastructure protection and quantum-resistant crypto.
- Intelligent approaches to eliminate security vulnerabilities in systems, services and applications: working on trusted supply chain for resilient services and security & privacy by design.

Fig. 4.1 BDVA strategic priorities (BDVA SRIA)

• From security components to security services: focusing on processes required to provide, manage, measure, certify and restore privacy and security.

Big Data Value Association (BDVA)

The BDVA is a non-profit organisation that implements the Big Data Value Private Public Partnership representing the private sector of this private-public collaboration. The main goals of the BDVA are to boost European Big Data Value research, development and innovation and to foster a positive perception of Big Data Value [8]. The BDVA plays a crucial role in the implementation of the Digital Single Market Strategy in Europe, contributing to its different pillars.

The mission of the BDVA is to develop the Innovation Ecosystem, that will enable the data-driven digital transformation in Europe, achieving and sustaining Europe's leadership in the fields of Big Data Value generation and Artificial Intelligence.

With its mission and vision in mind, the BDVA defines four different strategic priorities indispensable to achieve them. These priorities are shown in Fig. 4.1.

By its nature, Big Data can deeply influence cybersecurity and this relation emerges in its research and innovation efforts, in which the BDVA is involved in a number of topics:

• Homomorphic encryption
• Threat intelligence
• Assurance in gaining trust

- Differential privacy techniques for privacy-aware Big Data analytics
- Protection of data algorithms

The industry around data intelligence realizes the concerns that exist around security of the data being processed, and the possibilities to deliver new security services based on data processing and analytics. To this end, the BDVA has a specific task force working on data privacy and data protection.

Networked Software and Services Initiative (NESSI)

NESSI is the European Technology Platform that promotes software, services and data with the aim of helping to resolve European societal and economic challenges across all sectors such as manufacturing, transportation, healthcare and energy. NESSI's vision is to allow European businesses and citizens to stay competitive by creating new economic value with new service offerings. The main objectives of NESSI are:

- Develop strategies and business-focused analysis of R&I
- Mobilise industry and other stakeholders
- Share information and enable knowledge transfer

The digital world is growing fast due to the convergence of the network of services, things and data. Key enablers for this transformation are technologies and developments such as cloud computing, social media, data analytics and mobile communication.

The organisation defines several Research and Innovation challenges [9] classified by two criteria: Characteristics and Dimensions. The different challenges are defined in Table 4.3.

Table 4.3 NESSI research challenges

| | | Characteristics | | |
		Physical embedded distributed computing	Secure trustworthy sustainable computing	Smart cognitive computing
Dimensions	Digital infrastructures	Computing continuum	Dynamic policies and behaviour	Self-adaptable infrastructures
	Software Technologies	Full-stack software	Programming policies	Self-aware programs
	Data	Data everywhere	Built-in protection of assets and actors	Knowledge and insight for decisions
	Services	Software defined sectors	Regulation in societies of services	Digital business agents

European Organisation for Security (EOS)

EOS was created in 2007 by European private sector providers focusing in cyber-security. It is composed of members of all the different European countries from relevant domains such as ICT, energy, finance, transport, services, research, etc.

The main purpose of the organization is to provide a platform for collaborative work and sharing of best practices between industry, research centers and European institutions, more specifically in the cybersecurity domain. Therefore, the main goal is to contribute to the development of a harmonised European security market that satisfies political, societal and economic needs through European innovation.

EOS interacts with policy makers, EU institutions and agencies, member states, international and European organisations.

The task force of EOS has worked in identifying European needs and developed a report of recommendations for the upcoming FP9. The main objectives they identified are:

- Double the budget of FP9 and of the security chapter.
- Prioritize market strategies to reduce the research-to-market gap.
- Cybersecurity topics under FP9 should be more focused in innovation and end-to-end approach.
- Participation of SMEs should be incentivised.

EOS' activities are distributed in several working groups such as Security Screening and Detection Technologies, Integrated Border Security, Cybersecurity, or Security Research. Regarding cybersecurity research and innovation, the EOS' Cybersecurity Working Group includes the following objectives:

- Promoting the European cybersecurity and ICT security ecosystem.
- Collaborating with the European Commission and relevant stakeholders to promote R&I in cybersecurity.
- Fostering competitiveness and growth of the cybersecurity industry.
- Encouraging the implementation of European programs leading to the deployment of cybersecurity capabilities.
- Supporting the harmonization of EU Regulations.
- Setting up a stable regulatory environment with customers and administrations.

ECSEL Joint Undertaking

ECSEL is a Public-Private Partnership for Electronic Components and Systems that aims at keeping Europe at the forefront of technology development. ECSEL supports the European industry, SMEs and research organisations by co-financing several projects and launching Calls for Proposal for research, development and innovation projects. The organisation is composed of multiple members:

- Three associations: EPoSS,[1] AENEAS[2] and ARTEMIS[3]
- European Union (European Commission)
- Member States and Associated Countries to the Framework Program H2020.

The ECSEL SRA defines certain research and innovation priorities related with Cyber-Physical Systems, IoT and Industry 4.0 facilities. Below is a list of cybersecurity related ECSEL strategic priorities:

- Embedded software architecture modeling
- Security & privacy by design
- Security analysis and evaluation
- Safety and security patterns
- Harmonized and interoperable security evaluation & certification
- Formal methods for side fault analysis and fault attacks
- Fully homomorphic encryption: algorithms and attacks
- Simulation environments for security evaluations of components and systems
- Blockchain, post-quantum cryptography, quantum cryptography, fingerprinting
- New design tools: new security metrics, simulation techniques for safety and security analysis, verification and validation of security requirements
- Identity and access management
- Data protection
- Safe & secure execution platform
- Cryptography
- Trusted devices identities.

ARTEMIS Industry Association

The ARTEMIS Industry Association is a private partnership that represents multiple actors of the European Embedded Intelligent Systems Market, Internet of Things Market and Digital Platforms Market. The association continuously promotes the research and innovation priorities of the members to the European Commission and Public Authorities.

ARTEMIS Industry Association pursues the ambition to strengthen Europe's position in Embedded Intelligent Systems and other technologies related with IoT and Digital Platforms.

The strategy of ARTEMIS is based on the three following foundations [10]:

1. Represent Research and Innovation actors in Embedded Intelligent Systems in Europe
2. Disseminate and support ECSEL Public-Private Partnership and Horizon 2020 projects

---

[1]European Technology Platform on Smart Systems Integration, https://www.smart-systems-integration.org/.

[2]Association for European NanoElectronics ActivitieS, https://aeneas-office.org/.

[3]ARTEMIS Industry Association, https://artemis-ia.eu/.

3. Support ECSEL Joint Undertaking with the community, involving the different members of the organisation: Industry, SMEs, universities and research institutes.

In the Strategic Research Agenda (SRA), the association defines a number of technological challenges and opportunities, defined as the ARTEMIS Priority Targets:

- Safety-critical Secure Systems
- Virtual World
- Big Data/Data Analytics
- Systems of Systems
- Cloud Services
- Internet of Things
- Autonomous, Adaptive and Predictive Control
- Computing & Multicore

In their last SRA, the association created a matrix where the technological challenges or building blocks and the applications priorities are defined in Fig. 4.2.

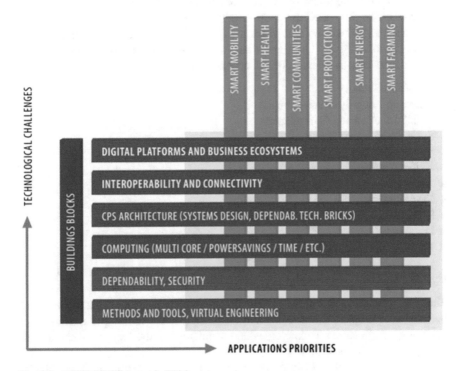

Fig. 4.2 ARTEMIS SRA matrix 2016

4.1.2.2   Industry Verticals Associations

Internet of Things (IoT)

The Alliance for Internet of Things Innovation (AIOTI) is a European association formed by IoT industrial companies, European research centers, universities and public bodies. The different efforts are performed through vertical Working Groups which focus on several defined areas of impact. These key areas are defined in 2 different levels:

- Horizontal: IoT Research, Innovation Ecosystems, IoT Standardisation, IoT Policy
- Vertical: Smart Living for Ageing Well, Smart Farming and Food Security, Wearables, Smart Cities, Smart Mobility, Smart Water Management, Smart Manufacturing, Smart Energy, Smart Building and Architecture.

The association identified some barriers that might restrict the take-up of IoT in the Digital Single Market related with privacy, security and liability. Due to these barriers, AIOTI is working on several topics such as: privacy, confidentiality of data, data authenticity, and integrity on IoT ecosystems.

According to the Baseline Security Recommendations for IoT report [11] , the deployment of IoT will be crucial for the European Smart cities and Smart infrastructures. In this context, the organization proposes some key technical areas to focus on in order to raise awareness for IoT cybersecurity:

- Hardware Security
- Trust and Integrity Management
- Data Protection and Compliance
- System Safety and Reliability
- Secure Software
- Authentication
- Access Control—Physical and Environmental Security
- Cryptography
- Secure and Trusted Communications
- Logging
- Monitoring and Auditing.

Communication Infrastructures

The 5G Infrastructure Public Private Partnership (5G-PPP) is a joint initiative between the European Commission and the European ICT Industry that will deliver solutions, architectures, technologies and standards for the coming generation of communication infrastructures.

The objective of the 5G-PPP is to secure Europe's leadership in the particular areas, where Europe has potential for creating new markets such as smart cities, e-health, intelligent transport, education or entertainment & media.

Key areas of research and innovation for 5G-PPP are related to the new 5G security architecture, defining the following challenges:

- A flexible architecture that integrates natively networking, computing and storage resources
- A secure and trustworthy network

  - Trusted and trustworthy 5G architecture
  - Multi-domain and multi-layer security
  - Security as a service

- A reliable and resilient network
- A quick end-to-end service deployment

In 5G-PPP, the 5G Infrastructure Association (5G IA) represents the private side while the European Commission represents the public side. The 5G IA brings together not only operators and manufacturers but also research institutes and universities. It also welcomes both verticals and SMEs. It hosts working groups dedicated to the advancement of 5G in Europe, and in particular a Security WG, which was created from a 5G-PPP Phase 1 project, 5G-ENSURE. The Security WG is not only open to 5G-PPP projects' participants but also to any 5G IA member interested in developed the 5G Security Community. The Security WG also publishes white papers and organises workshops, while developing liaisons with other security communities. The 5G IA is currently interested in:

- 5G threat landscape
- 5G slicing
- Access Control
- Security needs of verticals.

Transport and Mobility

ERTICO is an organization that brings together the interests of private and public European stakeholders. Its main objective is to bring intelligence into mobility for a safer, smarter and cleaner mobility by implementing deployment enablers and evaluating related technologies.

Against this background, the organisation focuses its efforts on several areas:

- Connected and Automated Driving
- Urban Mobility
- Transport & Logistics
- Cross-sectors
- Innovation Platforms

ERTICO is also undertaking some challenging topics such as Blockchain technologies, eCall, Cybersecurity and Privacy. According to the European Road Transport Research Advisory Council (ERTRAC), the ongoing European activities towards security regulation will affect connected and automated driving. The organization outlines core activities proposed by the industry:

- Cybersecurity Risk management tools across the value chain with specific focus on automation use cases

- Standardization of this approach in the ISO AWI 21434 "Road vehicles – Cyber-security Engineering"
- System Reliability and Security.

ERTRAC highlights the importance of data collection and data sharing for the different mobility stakeholders.

Manufacturing

European Factories of the Future Research Association (EFFRA) is a non-profit association that promotes the development of production technologies in the manufacturing industry. The partnership aims to bring together private and public resources to create an industry-led programme in research and innovation with the aim of launching market-oriented cross-border projects throughout the European Union.

The association defines several areas where the manufacturing technology innovation should focus its efforts. One of these areas is highly related with the cybersecurity topic, and it is defined as "Interoperable Digital Manufacturing Platforms: Connecting Manufacturing Services".

From a technological point of view, the IoT related technologies adoption will increase the efforts in data security and protection, due to the decentralization of the data which is now distributed in computing and cloud environments. Within this context, EFFRA highlighted the following cybersecurity needs:

- New Information Security approaches and related standards
- Identify Security Service Level Agreements
- Develop security and privacy frameworks in the new digitized context
- Accountability and Trust Management
- Models on data exchange and trade among 3rd party suppliers

The organization also identifies needs on several specific technological enablers:

- Concurrent Control Systems
- Models for data ownership
- Computation on encrypted data
- Critical data-management components
- Metadata structure for new data frameworks
- Data architecture for the data marketplace
- Machine learning algorithms.

Health

The European Federation for Medical Informatics (EFMI) is a non-profit organization which main objective is to research and develop medical informatics within the Health market. The association focuses its efforts through 14 different working groups. The one related with cybersecurity is the WG12: Security, Safety and Ethics.

Nowadays eHealth systems are increasingly subject to incidents that may affect their availability. Considering this fact, the EFMI association has positioned itself towards a research group investigating security and privacy tools. Their main topics of research are:

- Interoperable and trustworthy EHR[4] system solutions and personal health systems
- New healthcare and health management models for security and privacy
- Privacy and Security in the reuse and secondary use of health data
- Solutions for trustworthy and interoperable EHR and personal health systems

According to the MedTech Europe Association, which vision is aligned with the Global Medical Technology Alliance, safety is critical to the medical technology industry, and medical devices companies should take into account the critical need to assess security for their portfolio. The association exposes some principles that should be considered by the manufacturers when it comes to the production of medical technology devices:

- Medical device development and security risk management
- System-level security
- Coordinated disclosure
- Consensus standards, regulatory requirements and education.

4.1.2.3   Summary

As a summary and after the wide evaluation of topics and efforts of associations in Europe dealing with cybersecurity research & innovation, we can sum up several priorities related to cybersecurity, proposed by these organizations in their Strategic Research and Innovation Agendas (SRIA). These priorities are listed in the following summaries.

Summary of horizontal association interests

- ECSO

  - Data security and privacy technologies
  - Education and training
  - Security assessment and risk management
  - Distributed identity and trust management
  - User centric security and privacy
  - ICT Infrastructure Protection
  - Quantum-resistant cryptography
  - Trusted supply chain
  - Security and privacy by design

- BDVA

  - Privacy-aware big data analytics
  - Threat intelligence
  - Assurance in gaining trust
  - Protection of data algorithms
  - Homomorphic encryption

---

[4]Electronic Health Record.

- NESSI

  - Data Everywhere
  - Dynamic policies and behavior
  - Built-in protection of assets and actors
  - Regulation in societies of services

- ECSEL

  - Security & privacy by design
  - Fully Homomorphic Encryption
  - Cryptography
  - Identity and access management
  - Data protection
  - Safety and security patterns
  - Trusted devices identities

Summary of vertical industries interests

- Internet of Things

  - Hardware and software security
  - Trust and integrity management
  - Cryptography
  - Data protection and compliance
  - Authentication
  - Monitoring and auditing

- Communication Infrastructures

  - Trusted 5G architecture
  - Multi-domain and multi-layer security
  - Security as a service

- Security

  - Policy and EU policy regulation harmonization

- Transport and Mobility

  - Blockchain technologies for security
  - Privacy
  - eCall
  - Risk management
  - Reliability and security
  - Standardization

- Manufacturing

  - Security and privacy frameworks
  - Trust management

- Cryptography
- Machine learning
- Critical data management
- Data exchange and trade

• Health

- Privacy and security
- Security risk management
- Trustworthy EHR system solutions

All the analysed associations have a common interest in making their industries secure markets, and all of them dedicate great efforts to promote research and innovation of cybersecurity solutions. This fact implies that cybersecurity is not just a common interest for the industries, but a requirement.

We see several common topics of interest across industries and associations: privacy, cryptography and risk management are referenced in several of these European-level groups.

4.1.3 Associations and Clusters in Japan

There are a number of industry associations that influence research activities in Japan through reports, policy dialogue, representation at government subcommittees, as well as deliberations through working groups. In this section, we highlight key industry associations that are related with the ICT sector, with considerable capabilities to generate potential impact on the cybersecurity market in Japan.

Acronym	Name	Main industry
Keidanren	Japan Business Federation	All industries
JNSA	Japan Network Security Association	Network security
VLED	Vitalizing Local Economy Organization by Open Data & Big data	Open data & Big data
5GMF	Fifth Generation Mobile Communication Promotion Forum	5G
IoTAC	IoT Acceleration Consortium	IoT

4.1.3.1 Japan Business Federation (Keidanren)

Keidanren (Japan Business Federation) is a comprehensive economic organization with a membership comprised of 1,370 representative companies of Japan, 109 nationwide industrial associations and 47 regional economic organizations (as of April 1, 2018).

Its mission as a comprehensive economic organization is to draw upon the vitality of corporations, individuals and local communities to support corporate activities which contribute to the self-sustaining development of the Japanese economy and improvement in the quality of life for the Japanese people.

For this purpose, Keidanren establishes consensus in the business community on a variety of important domestic and international issues for their steady and prompt resolution. At the same time, it communicates with a wide range of stakeholders including political leaders, administrators, labor unions and citizens.

Keidanren implements diverse policy activities through its broad range of policy committees that also cover cybersecurity issues. It collectively contributes to the public-private policy dialogue through policy proposal documents as well as its representation at government committees. Regarding innovation, a number of policy committees exist, such as the committee on information and telecommunication, committee on new industry and technology, and the committee on new business and medium enterprises.

Keidanren has published several policy proposals in the area of cybersecurity:

- A Call for Reinforcement of Cybersecurity to Realize Society 5.0 (December 12, 2017)
- Second Proposal for Reinforcing Cybersecurity Measures (January 19, 2016)
- Proposal for Reinforcing Cybersecurity Measures (February 17, 2015)
- Toward the Establishment of a Safe, Secure Internet Society through International Cooperation (October 17, 2006)

Furthermore, Keidanren has revised its Charter of Corporate Behavior on November 8, 2017 as part of its recent activities to embrace digital transformation and cybersecurity. Its Implementation Guidance on Charter of Corporate Behavior specifically invites members to "strive to ensure cyber security" through organizational planning, risk management, resource allocation, information sharing and incident readiness.

4.1.3.2   Japan Network Security Association (JNSA)

The Japan Network Security Association is a non-profit organisation with industry membership, where different working groups develop different facets of the Association's charter.

The objective of the JNSA is to promote the standardization of the network security technologies, contributing to better standards and measures in the field, and enhancing the public welfare through research and dissemination activities within the Network Security market.

JNSA is an industry association encompassing broad range of issues, ranging from awareness, education, membership services and investigation of standards. It has more than 20 working groups under 5 committees: social activities, survey and research, standards investigation, education, and marketing.

JNSA published a number of reports, checklists, guidelines and textbooks in Japanese. For instance, the following documents were published in 2017:

- Six Ws on cybersecurity information sharing for enhancing SOC/CSIRT (Oct 30, 2017)
- Textbook for security incident response organization 2.0 (Oct 3, 2017)
- Security Body of Knowledge (SecBoK) skill map 2017 (Aug 21, 2017)
- Checklist for the introduction of identity management systems (Jul 12, 2017)
- Japan IT security market analysis report 2016 (Jun 20, 2017)
- Information security incidents in 2016 (Jun 14, 2017)
- Web application vulnerability assessment guideline (Apr 24, 2017)

4.1.3.3 Vitalizing Local Economy Organization by Open Data & Big Data (VLED)

"Vitalizing Local Economy Organization by Open Data & Big Data" is an organization established with views to promote the public opening of the data owned by public service (public sector) entities, and to advance local vitalization (regional revitalization) and activation of economies, through utilization of such data in combination with the data owned by the national government and local public bodies, as big data.

4.1.3.4 Fifth Generation Mobile Communication Promotion Forum (5GMF)

The objectives of the Fifth Generation Mobile Communication Promotion Forum (5GMF) are to conduct research & development concerning Fifth Generation Mobile Communications Systems, research and study pertaining to standardization thereof, along with acting as a liaison and coordinating with related organizations, collecting information, and promoting education and awareness about Fifth Generation Mobile Communications Systems aimed at the early realization thereof, all with the aim of thereby contributing to the sound development of the use of telecommunications.

4.1.3.5 IoT Acceleration Consortium (IoTAC)

The IoT Acceleration Consortium was established in Japan with the aim of creating an adequate environment for attracting investment in the future with the Internet of Things (IoT) through public-private collaboration. The background is the possibility of a great change in conventional industry and structure of society with the development of IoT, Big Data, and artificial intelligence (AI) in recent years. While the development of IoT creates new services using data, it also generates concerns about existing businesses rapidly becoming obsolete.

The IoT Acceleration Consortium aims to combine the strengths of government, industry, and academia and build a structure for developing and demonstrating technologies related to the promotion of IoT as well as creating and facilitating new business models. The consortium promotes (1) the development, demonstration, and standardization for IoT-related technologies and (2) creation of various IoT-related projects and recommendations such as regulatory reform to enable those projects.

4.1.3.6   Society 5.0: National Strategic Effort in Japan

Society 5.0 was proposed in the 5th Science and Technology Basic Plan as a future society that Japan should aspire to: "a human-centered society that balances economic advancement with the resolution of social problems by a system that highly integrates cyberspace and physical space." The term was introduced to inform broadest stakeholders that the next wave of industrial revolution is coming, after the hunting society (Society 1.0), agricultural society (Society 2.0), industrial society (Society 3.0), and information society (Society 4.0) [12].

Since its introduction, the term has been widely embraced by both industries and government agencies as the generational conceptualization has been useful to signal industries that the next wave of innovation is going to transform our societies. Society 5.0 also implies that such next wave of innovation should foster balanced economic development, tackle societal challenges such as sustainable development, and nurture human-centered society.

Society 5.0 serves as a common thread to connect industry associations as it is driven by commonly accepted values, while at the same time proposing to incorporate enabling technologies such as IoT, big data and cybersecurity into people's daily lives. Major industry associations thus adopt the term, in order to engage in the industry and policy dialogue to explore variety of innovation pathways toward Society 5.0.

4.1.4   Associations and Initiatives at Member States level

4.1.4.1   Spain

Overview provided in Table 4.4.

Agrupación Empresarial Innovadora Ciberseguridad y Tecnologías Avanzadas (AEI)

The AEI is an organisation that brings together universities, research centers, companies and other organisations interested in promoting new technologies. The AEI is part of the organ of government of the European Cyber Security Organisation (ECSO)[5] The AEI performs a large number of roles, such as:

---

[5]Agrupación Empresarial Innovadora Ciberseguridad y Tecnologías Avanzadas, aeiciberseguridad.es.

Table 4.4 Associations and initiatives in Spain

Acronym	Name	Main industry
AEI ciberseguridad	The Spanish Cybersecurity Innovation Cluster	Cybersecurity
INCIBE	Instituto Nacional de Ciberseguridad de España	Cybersecurity
ISMS Forum Spain	Asociación Española para el Fomento de la Seguridad de la Información	Information Security
CCI	Centro de Ciberseguridad Industrial	Cybersecurity

- Promoting business and technical services implementation by engaging a dialogue between R&D entities, companies and professionals.
- Promoting new strategies for security of information technologies and cybersecurity.
- Developing collaborative exchange in innovative ICT industry projects execution.

AEI defines the following areas of strategic development related with cybersecurity technologies:

- Anti fraud
- Antimalware
- Technical and forensic audit
- Digital Authentication
- Firewall
- Event Management
- Visual recognition access and identity control
- Security on mobility
- Cryptographic systems

Instituto Nacional de Seguridad de España (INCIBE)

INCIBE is a Spanish organism founded in 2006 and dependent on the Energy, Turism and Digital Agenda Ministry. Its main objective is to develop the Information Society through innovation and Cybersecurity project development with international view.[6] INCIBE's activity is founded on three fundamental pillars:

- Services: The organisation works for the user's protection and privacy by promoting mechanisms for the prevention of data security incidents
- Research: by addressing a wide range of complex projects with an innovative nature
- Coordination: participating in multiple partnership networks, giving the organisation immediate capabilities to execute operations in the field of cybersecurity.

---

[6]Instituto Nacional de Ciberseguridad, www.incibe.es.

The INCIBE defines 12 research and innovation strategic topics:

- Reliable and updatable systems
- Privacy
- Data processing
- Critical infrastructures
- Systems evaluation and Cyber Risks
- Attacks and defense against threats
- Identity Management
- Cryptography
- Safety awareness promotion
- Network Security
- Cloud Computing
- Internet of Things

Asociación Española para el Fomento de la Seguridad de la Información (ISMS)

The ISMS is a non-profit organisation founded in 2007 aimed at promoting the development, knowledge and culture of Information Security in Spain and benefiting the whole community within this sector. The association implements its activities through different initiatives:

- International journeys
- Data Privacy institute
- Cloud Security Alliance
- Cyber Security Center

The ISMS works in 5 different areas: Cybersecurity, Privacy, Mobility, Cloud, and Training. The Cybersecurity center promotes the exchange of knowledge between the key actors and experts in the sector to boost and contribute to the cybersecurity improvements in Spain.

Centro de Ciberseguridad industrial (CCI)

The Center of Industrial Cybersecurity is a non-profit organisation founded in 2013 with the mission of contributing to the continuous improvement of Industrial Cybersecurity in Spain and Latin America.

The CCI does not define research trends or strategic topics, so any topic regarding or including industrial cybersecurity is of its interest.

4.1.4.2 France

Overview provided in Table 4.5.

Systematic Paris Region Digital Ecosystem (SPR)

Systematic is a Global Competitiveness Cluster that brings together more than 800 members connecting software, digital and industry players. Its main objective is to

Table 4.5 Associations and initiatives in France

Acronym	Name	Main industry
Systematic	Systematic Paris Region Digital Ecosystem	Multiple Sectors
FIC	The FIC observatoire	Cybersecurity
ALLISTENE	L'alliance des sciences et technologies du numérique	ICT
CLUSIF	The French Cybersecurity Club	Multiple Sectors
Hexatrust	Cloud Confidence and Cybersecurity	Multiple Sectors

accelerate digital project through collaborative innovation, SME development and networking in several sectors: energy, ICT, Health, Transport, etc. For Systematic Ecosystem, the innovation for these markets is based on the development and continuous growth of the following key technologies:

- Digital Confidence & Cybersecurity
- Big Data & AI
- Digital infrastructures
- Free software
- Modeling, numerical simulation and HPC
- Embedded Systems & Internet of Things.

FIC Observatoire (FIC)

The FIC Observatoire is a platform co-hosted by the national French Gendarmerie for debates and exchanges about cybersecurity and digital trust. The FIC aims at establishing a common framework for converging initiatives around supply and demand with two objectives:

- Understand and identify the needs of users
- Offer a prospective vision of the industrial sector

L'alliance des sciences et technologies du numérique (ALLISTENE)

Allistene supports the economic and social changes related to the diffusion of digital technologies. The organisation's objective is to coordinate the various players in digital science and technology research in order to develop a coherent and ambitious research and technological development program. The mission is to identify common scientific and technological priorities and to strengthen partnerships between public operators (universities, schools, institutes), while creating new synergies with companies.

Club de la sécurité de l'information français (CLUSIF)

CLUSIF promotes the exchange of ideas and experiences through working groups, publications and conferences. The club keeps watch of the trending cybersecurity topics as well as the latest vulnerabilities and breaches. Their main objective is to control exposure to risk in general, and in information systems in particular. The club is articulated around two main communities: the "Providers", i.e., the security solutions and services vendors, and the "Users", i.e., the corporate CISOs. Latest key topics that are discussed include:

- the NIS directive
- the Internet of Things
- the Blockchain
- the GDPR
- Industrial Control Systems
- incident management
- DevOps.

Hexatrust

Hexatrust gathers innovative companies from both the Cloud Computing and the cybersecurity sectors. Hexatrust promotes cloud confidence and cybersecurity through a "one-stop-shop" offering certified security solutions to fulfil the needs of both public and private companies, including:

- regulatory compliance: GDPR, eiDAS, NIS, LPM (the French Military Legal Program)
- protection against cyber-threats
- support in digital transformation projects.

4.1.4.3 Poland

Overview provided in Table 4.6.

Fundacja Bezpieczna Cyberprzestrzeń / Cybersecurity Foundation (CSF)

The Cybersecurity foundation is a non-governmental organization established in June 2010. The main objectives of the foundations are to educate and conduct cyber exercises and develop activities of research related to IT security and critical infrastructure protection. Its mission is to raise awareness on threats emerging from the cyber domain.

One of the most valuable initiatives of the foundation is the Cyber-EXE^{TM} Polska, aimed at exploring the capacity and preparation of their members to identify and address the different threats in the field of ICT Security.

Table 4.6 Associations and initiatives in Poland

Acronym	Name	Main industry
CSF	Fundacja Bezpieczna Cyberprzestrzeń/Cybersecurity foundation	Cybersecurity
–	Mazowiecki Klaster ICT	ICT

Table 4.7 Associations and initiatives in Greece

Acronym	Name	Main industry
GCC	Greek Cybercrime Center	Cybersecurity

Mazowiecki Klaster ICT

The Mazovia Cluster ICT is a cluster established in 2007 and coordinated by the Association of Economic and Social Development "knowledge".[7] The mission of the Mazovia Cluster ICT is to bridge the gap between action and conditions for SME's development by establishing and developing cooperation between sectors of computer science enterprises, telecommunication and electronic media R&D units, universities and business institutions.

Regarding cybersecurity and security aspects of ICT systems, the Cluster points out several challenges and areas of research:

- Public Key Infrastructure systems
- Safe systems of document circulation
- Identity Management systems
- Safety audits of IT systems
- Analysis and expertise in information security, electronic documents, identity cards, PKI.

4.1.4.4   Greece

There is not a wide range of Greek associations and clusters specialized in Cyber-security. However there are existing projects and initiatives aimed at mobilizing the Greek constituency in the area of cybercrime training, research, education and uptake of results (Table 4.7).

Greek Cybercrime Center

The Greek Cybercrime Center is part of an European effort that has received funding from the Prevention of and Fight against Crime Programme of the European Commission.

---

[7]Mazowiecki Klaster ICT, http://klasterict.pl/about-us/.

Table 4.8 Associations and initiatives in Belgium

Acronym	Name	Main industry
CSC	Cyber Security Coalition	Cybersecurity
BELTUG	Beltug	Digital technology

This initiative combines partners from the industry, academia and law enforcement that together aim to promote the state-of-the-art in cybercrime.

4.1.4.5   Belgium

Overview provided in Table 4.8.

Cyber Security Coalition (CSC)

The Cyber Security Coalition is a partnership between entities from the academic world, private sector and public authorities from Belgium that join forces to fight against cybercrime. The coalition mission focuses on generating a cross-sector collaboration with various objectives[8]:

- Share knowledge and experience
- Initiate, organize and coordinate concrete cross-sector initiatives
- Raise awareness among citizens and organizations
- Promote the development of expertise
- Issue recommendations for more efficient policies and regulations

To approach this mission, the cluster is organized in 4 strategic axes, with their own working groups.

- Experience Sharing. Trusted Platform to exchange knowledge and information on specific Cybersecurity issues. With specific themes distributed in multiple sessions: GDPR, Cyber heists, Cyber Ranges, etc…
- Operational Collaboration: Collective actions to fight cybercrime. Topics of discussion in this working group are:

  - Response & technical tooling to support incident response
  - Detection, future-proof monitoring & use cases
  - Working transformation, Cloud & Mobile Devices
  - Privileged Access Management, GDPR
  - Ransomware, awareness exercises, security by design
  - Threat-hunting activities
  - SSL interception
  - Wannacry-like incidents
  - PSD2 & API's

---

[8] The Cyber Security Coalition, https://www.cybersecuritycoalition.be.

- Policies & Recommendations: Support sectorial bodies in setting policies and defining new ways to implement these policies.
- Awareness Raising: Dissemination campaigns, events and conferences.

Belgian Association of Digital Technology Leaders (BELTUG)

Beltug is the largest Belgian association of digital technology providers. It is focused on breaking down the barriers that the companies face when building their digital strategy.

The association prioritizes several topics related with the digital technology market, collected in their action plan:

- Blockchain
- Smart collaboration
- Digital Signature (eIDAS)
- IoT Security
- Enterprise mobility management
- Workable Security
- Managed Security Services
- Hybrid IT integration in ICT architecture
- 5G
- GDPR: Cloud Vendor assessment, HR Services assessment, Template consent, Special Interest Group, Corporate policy and privacy.
- e-archiving
- e-privacy
- CAPEX & Cloud
- Software CoC
- Access National Number.

4.2 Common Topics of Interest

From this analysis, we can conclude that there are indeed common topics of interest around cybersecurity between the Japanese and European industry needs.

Industry in both regions has identified a strong interest in areas like 5G and big data, with specific organizations created in both regions to promote these technologies and to address its specific cybersecurity challenges in these areas. More specifically, we can find BDVA and 5GPPP in Europe, while VLED and 5GMF were created in Japan with similar objectives. There exist also IoT efforts in both regions in the form of associations.

The Japan Network Security Association has also similar counterparts in some European countries, although a EU-level equivalent does not exist as such. Some of the EU-level organizations (such as ECSO) address some of the similar objectives as JNSA, to some extent.

Regarding SMEs, we have observed how both Europe and Japan have identified as key priorities to involve them more in research, development and innovation programs so they can on the one hand provide their needs and expertise and on the other hand benefit from the cybersecurity activities.

In conclusion, we can see that several key areas of industry-led research around cybersecurity that are common in both regions can be found in applications and technologies around Big Data, 5G and IoT.

Common research interests include privacy of big data, availability and reliability of open data, security of 5G communication networks and protocols, cryptography, and regulatory aspects of cybersecurity to enable new applications and technology developments.

References

1. 43% of cyberattacks target small businesses (2018). https://www.prnewswire.com/news-releases/43-of-cyberattacks-target-small-businesses-300729384.html. Cited 15 Nov 2019
2. Quicke, S.: Reports underline SME exposure to cyber attacks (2018). https://www.computerweekly.com/microscope/news/252462240/Reports-underline-SME-exposure-to-cyber-attacks. Cited 15 Nov 2019
3. Ismail, N.: Ransomware, Emotet and Trojan attacks against businesses on the rise—Malwarebytes cybercrime report (2019). https://www.information-age.com/cyber-attacks-malwarebytes-cybercrime-report-123481980/. Cited 15 Nov 2019
4. ENISA: Industry 4.0—Cybersecurity challenges and recommendations (2019). https://www.enisa.europa.eu/publications/industry-4-0-cybersecurity-challenges-and-recommendations. Cited 15 Nov 2019
5. Sidat, L.: A significant number of businesses are still not investing enough in cyber security (2019). https://www.cheshire-live.co.uk/special-features/businesses-not-investing-cyber-security-16180726. Cited 15 Nov 2019
6. Biggs, J.: Study finds most ransomware solutions just pay out crypto (2019). https://www.coindesk.com/study-finds-most-ransomware-solutions-just-pay-out-crypto. Cited 15 Nov 2019
7. ECSO: Strategic research and innovation agenda (2017). https://www.ecs-org.eu/documents/publications/59e615c9dd8f1.pdf. Cited 1 Dec 2019
8. Big data value association: European big data value strategic research and innovation agenda (2017). http://www.bdva.eu/sites/default/files/BDVA_SRIA_v4_Ed1.1.pdf. Cited 1 Dec 2019
9. NESSI: Strategic research innovation agenda (2017). http://www.nessi-europe.com/files/NESSI_SRIA_2017_issue_1.pdf. Cited 1 Dec 2019
10. ARTEMIS Industry Association: Artemis strategic research agenda (2016). https://artemis-ia.eu/publication/download/sra2016.pdf. Cited 1 Dec 2019
11. ENISA: Baseline security recommendations for IoT (2017). https://www.enisa.europa.eu/publications/baseline-security-recommendations-for-iot. Cited 1 Dec 2019
12. Cabinet Office: Society 5.0 (2019). http://www8.cao.go.jp/cstp/english/society5_0/index.htm. Cited 1 Dec 2019

Chapter 5
Summary

There is no doubt that synchronization of efforts in many areas in the field of cyber-security is very important. Harmonization or at least mutual understanding of the terms of individual regulations can bring about many synergies.

In the book, the authors have analyzed cybersecurity priorities in both the EU and Japan, in order to produce an overview on the status and priorities of research and innovation in cybersecurity and privacy in Europe and Japan. The main findings can be divided (like the analysis described in the book) into three fields: law and policy, research and innovation, and industry and stardardization.

A legal and regulatory landscape have been elicited, with particular attention to the GDPR and focusing on the analysis of regulatory documents, EU directives, as well as laws and legal frameworks. The book presents common regulatory aspects of cybersecurity and privacy, for example obligations regarding monitoring, certi-fication, protection of personal data and information exchange. Both regions have undergone significant reforms and modernization of the existing legal frameworks with regard to privacy. In Europe, the GDPR brings up a significant update to leg-islation with a strict compliance bar foreseen for its enforcement. In Japan, the Act on Personal Information was amended. However, in spite of mutual recognition, it should be noted that these two frameworks are not perfectly matched. For example, the concepts of sensitive personal data, as well as some practical implications, such as a divergent approach to imposing fines and sanctions might become a critical point for both Japanese and European companies and organizations wanting to enter each other's digital markets. In contrast, the cybersecurity domain is slightly different than the one described above. At first glance, there are differences in the regula-tions of both regions, mainly in terms of their scope and the stakeholders affected by these provisions, thus diverging in substantial elements. However, some lines of similarities can be pointed out. This can be observed in the room left by both policy and legal frameworks enabling EU, Member States and Japanese Government to engage in international cooperation, particularly in the fields of international norms settings and common strategy building, promoting and raising awareness about the cybersecurity culture in relevant territories and not only.

A. Felkner et al., *Cybersecurity Research Analysis Report for Europe and Japan*, Studies in Big Data 75, https://doi.org/10.1007/978-3-030-62312-8_5

The book also focuses on analyzing of research agendas, programs, project calls and roadmapping projects as well as showing the available mechanisms to finance cybersecurity research. Some of the issues described include the financial mechanisms for research and innovation in cybersecurity in both regions, and the current role and activity of different units (SMEs, research institutions, CSIRTs, LEAs, etc.) in research and innovation. An overview of the main research directions in this field is also provided, in line with the identification of the strong and weak points in both regions in order to identify topics of common interest in which cooperation opportunities exist, and topics on which some aspects are covered asymmetrically. An analyzis of long-term research programs at national and international level is also provided, in order to find thematic parallels between the EU and Japan, which may create opportunities for either co-financed joint EU-Japan projects, or at least synchronization of efforts enabling cooperation. The conclusions have been prepared on the basis of many sources of information. First of all, the outcomes of questionnaires collected during and after the *EUNITY* project workshops have been deeply analyzed. Secondly, the observations and experience of all partners of the *EUNITY* consortium have been used to expand the landscape of the problems indicated in questionnaires and highlight typical problems. Another very important source of information is a review of strategies, projects and programs with regard to research and innovation in the area of cybersecurity and privacy, as well as relevant funding mechanisms. It is worth noting that in many aspects of research in cybersecurity and privacy, the strong and weak points are common for both regions. To sum up, the main strengths are: establishing a cybersecurity strategy and a declaration of concentration on cyber security and privacy. Whereas the main weaknesses are the following: opposition between industry and research, a high-level cybersecurity personnel shortage, a lack of coordination of research actions at various levels, and a lack of strong global cybersecurity enterprises and solutions originating in the EU and Japan. It is certainly very important to increase Europe's response as a whole and to support cooperation and coordination in the field of cybersecurity between Member States and Japan. There is a need of industry-government cooperation and global collaboration to exchange sensitive data and to enlarge the cooperation to as many countries and industry sectors as possible. Global collaboration shall not only be horizontal, i.e., limited to state entities, nations and international organizations. Rather, global cooperation should be horizontal and vertical, i.e. also involving private entities and other stakeholders (for example, the academic community).

In relation to industry and standardization aspects, the book presents the analysis aimed at understanding the efforts and measures undertaken and the challenges identified when it comes to cybersecurity. The analyzed initiatives have been chosen considering several criteria: scope of action (cross-cutting technologies and vertical markets), impact and activity level. Several of organizations were established to manage different contractual public-private partnerships (cPPP) on different topics. The cPPP is an instrument co-funded by the European Union for research and innovation activities in crucial sectors of the European economy, bringing together companies, universities, research laboratories, SMEs and other organizations. These cPPPs develop strategic research and innovation agendas in which the R&I priorities and challenges of the industry are structured and defined. Industry in both regions has identified a strong interest in areas such as 5G and big data, and in both areas special organizations have been created to promote these technologies and solve cybersecurity problems in these areas. BDVA and 5GPPP appeared in the EU, while VLED and 5GMF were created in Japan with similar goals. In both regions, activities related to the Internet of Things in the form of associations are undertaken. The Japan Network Security Association also has similar counterparts in some European countries, although an EU-level equivalent does not exist as such. Some EU-level organizations (such as ECSO) address some of the similar objectives as JNSA to some extent. In conclusion, we can see that several key areas of industry-led research around cybersecurity that are common in both regions can be found in applications and technologies around Big Data, 5G and the Internet of Things. Common research interests include the privacy of big data, the availability and reliability of open data, security of 5G communication networks and protocols, cryptography, and regulatory aspects of cybersecurity to enable the development of new applications and technology.

The authors hope that this publication can be seen as a small but necessary step towards increasing the effectiveness of cybersecurity activities in both the EU and Japan, and consequently changing our cyberworld to a safer one.

Appendix
National Cybersecurity Strategies

Brief Description of the Methodology of Cybersecurity Strategies Analysis

Strategies of EU Member States are analysed briefly in this chapter. National strategies of selected countries are analysed in a more detailed way, whereas analyses of strategies of other countries are based on information provided by ENISA on *National Cyber Security Strategies (NCSSs) Map* [1]. The description presents the state as at the date of the end of June 2018.

The most important questions to be answered on the basis of these analyses are defined as follows:

1. What research and innovation actions or objectives were defined in a national strategy?
2. What areas of research and innovation are defined in a strategy as especially important?
3. What are the goals beyond the strategy objectives defined by the ENISA (and briefly described below)?

In all of these questions we focus mainly on the research and innovation aspects. To enable making conclusions we briefly state the methodology behind these analyses and descriptions.

ENISA supports EU Member States in developing, implementing and evaluating their National Cyber Security Strategies (NCSS). Currently all Member States have developed the NCSS, but the scope of the documents, as well as their actuality, are different.

A. Felkner et al., *Cybersecurity Research Analysis Report for Europe and Japan*,
Studies in Big Data 75, https://doi.org/10.1007/978-3-030-62312-8

Objectives

ENISA has defined objectives and presented analysis of all countries' strategies in a form of list of objectives which are met by each of the strategies (Table A.1).

The ENISA NCSS Map lists all the documents of National Cyber Security Strategies in the EU together with their strategic objectives and good examples of implementation. The information is based on publicly available material. More detailed information about the objectives and good practices is based on the NCSS Good Practice Guide that was published by ENISA in 2016 [1].

In the context of the national strategies' analysis, *National Cyber Security Strategies: Practical Guide on Development and Execution* [2] and a newer document: *National Cyber Security Strategies: Good Practical Guide* [3], both prepared by ENISA, are very important and helpful. These documents, among other things, define the areas of interest of a cyber-security strategy. These areas are described in the documents mentioned above and can be briefly summarized as follows.

In the context of research and innovation, the most important objective is "Foster R&D", the more detailed description of this objective, on the basis of the document *National Cyber Security Strategies: PracticalGuide on Development and Execution* [2], is as follows: *Typical objectives of this phase include the following.*

- *Identify the real causes of the vulnerabilities instead of repairing their impact.*
- *Bring together scientists from different disciplines to provide solutions to multidimensional and complex problems such as physical-cyber threats.*
- *Bring together the needs of industry and the findings of research, thus facilitating the transition from theory to practice.*
- *Find ways not only to maintain but also to increase the level of the public trust in existing cyber infrastructure.*

Typical tasks in this step comprise the following:

- *Create a forum for industry and request R&D topics for consideration.*
- *Create an R&D agenda with topics that support the objectives of the cyber security strategy with a midterm horizon, of between four and seven years. For each topic the following should, at least, be described: the objective, the incentives and the challenges of the topic.*
- *Create a platform for bringing together high-level research and the private sector. This platform may take the form of a public/private partnership. Seek cooperation with similar European and international relevant activities i.e. the European Commission Research and Innovation programme (e.g. FP7/8).*
- *Create a coordination research plan in order to avoid overlaps between research activities undertaken by different institutions and programmes.*
- *Develop effective incentives to make cyber-security research ubiquitous. Both individuals and organisations might be the beneficiary of these incentives.*

Table A.1 Objectives which should be defined in National Cybersecurity Strategies (NCSSs)

Objective	Description (based on ENISA's NCSS Good Practice Guide [3])
Set the vision, scope, objectives and priorities	*The aim of a cyber security strategy is to increase the global resilience and security of national ICT assets, which supports critical functions of the state or of the society as a whole. Setting clear objectives and priorities is thus of paramount importance for successfully reaching this aim*
Follow a national risk assessment approach	*One of the key elements of a cyber-security strategy is the national risk assessment, with a specific focus on critical information infrastructures. Risk assessment is a scientific and technologically based process consisting of three steps: risk identification, risk analysis and risk evaluation*
Take stock of existing policies, regulations and capabilities	*Before defining in detail the objective of the cyber-security strategy, it is important to take stock of the status of the key elements of the strategy at national level. At the end of this activity important gaps must be identified*
Develop a clear governance structure	*A governance framework defines the roles, responsibilities and accountability of all relevant stakeholders. It provides a framework for dialogue and coordination of various activities undertaken in the lifecycle of the strategy. A public body or an interagency/interministerial working group should be defined as the coordinator of the strategy*
Identify and engage stakeholders	*Identifying and engaging stakeholders are crucial steps for the success of the strategy. Public stakeholders usually have a policy, regulatory and operational mandate. They ensure the safety and security of the national critical infrastructures and services. Selected private entities should be part of the development process due to the fact that they are likely the owners of most of the critical information infrastructures and services*
Establish trusted information-sharing mechanisms	*Information-sharing is a form of strategic partnership among key public and private stakeholders. Owners of critical infrastructures could potentially share with public authorities their input on mitigating emerging risks, threats, and vulnerabilities while public stakeholders could provide on a 'need to know basis' information on aspects related to the status of national security, including findings based on information collected by intelligence and cyber-crime units*

(continued)

Table A.1 (continued)

Objective	Description (based on ENISA's NCSS Good Practice Guide [3])
Develop national cyber contingency plans	*A national cyber security contingency plan should be part of an overall national contingency plan. It is also an integral part of the cyber security strategy*
Protect critical information infrastructure	*Critical information infrastructure protection (CIIP) is an integral part of many cyber and information security strategies. Cybersecurity covers a broad spectrum of ICT-related security issues, of which the protection of CII is an essential part*
Organise cyber security exercises	*Exercises enable competent authorities to test existing emergency plans, target specific weaknesses, increase cooperation between different sectors, identify interdependencies, stimulate improvements in continuity planning, and generate a culture of cooperative effort to boost resilience*
Establish baseline security requirements	*Baseline security requirements for a given sector define the minimum security level that all organisations in that sector should comply with. Such requirements could be based on existing security standards or frameworks and good practices widely recognised by the industry*
Establish incident reporting mechanisms	*Reporting security incidents plays an important role in enhancing national cyber security. Incident reporting and analysis helps in adjusting and tailoring the security measures list, referred to in the previous section, to the changing threat landscape*
Raise user awareness	*Raising awareness about cyber-security threats and vulnerabilities and their impact on society has become vital. Through awareness-raising, individual and corporate users can learn how to behave in the online world and protect themselves from typical risks. Awareness activities occur on an ongoing basis and use a variety of delivery methods to reach broad audiences*
Strengthen training and educational programmes	*Unfortunately, universities and R&D institutions do not produce enough cyber-security experts to meet the increasing needs of this sector. Cyber security is usually not a separate academic topic but part of the computer science curriculum. Cyber security is also a continuously changing topic that requires constant training and education*

(continued)

Table A.1 (continued)

Objective	Description (based on ENISA's NCSS Good Practice Guide [3])
Establish an incident response capability	*National/governmental CERTs play a key role in coordinating incident management with the relevant stakeholders at national level. In addition, they bear responsibility for cooperation with the national/governmental teams in other countries*
Address cyber crime	*The fight against cyber crime requires the collaboration of many actors and communities to be successful. In this respect, it is important to address and counter the rise of cyber crime and to prepare a concerted and coordinated response with relevant stakeholders*
Engage in international cooperation	*Cyber security threats and vulnerabilities are international in nature. Engaging in cooperation and information sharing with partners abroad is important to better understand and respond to a constantly changing threat environment*
Establish a public/private partnership	*A public/private partnership (PPP) establishes a common scope and objectives and uses defined roles and work methodology to achieve shared goals. PPPs may focus on different aspects of security and resilience*
Balance security with privacy and data protection	*Counter-terrorism measures and tools that tackle cyber crime often invade privacy in the most brutal ways and, at the same time, lack of personal online security leads to breaches of that same privacy. A cyber-security strategy should seek for the right balance between these two concepts*
Institutionalise cooperation between public agencies	*Cyber security is a problem area that spans across different sectors and across the responsibilities of different public agencies. Therefore, close cooperation between these entities is an important pillar for the successful implementation of NCSS*
Foster R&D	*Research and development in cyber security is needed in order to develop new tools for deterring, protecting, detecting, and adapting to and against new kinds of cyber attacks*
Provide incentives for the private sector to invest in security measures	*There are different ways how governments can ensure that businesses implement appropriate security measures. One way is to make certain standards mandatory by law. However, governments can also apply softer steering measures, for example by giving incentives to businesses to invest in certain security measures*

The EU Member States' National Cybersecurity Strategies

The national strategies of cybersecurity should, among other things, state the main research directions pointed out by the countries. Because of that, in this section there is a need to briefly describe national strategies. The strategies of five selected countries (namely: Frace, Belgium, Greece, Poland and Spain) are described in more detail, as an example, the strategies of the rest of the EU' countries are just mentioned.

France

The version of the law described for this deliverable covers the 2019–2025 period, with the objective of preparing 2030. As such, there are very few changes at this stage, except that the law has been officially adopted by the French parliament and published by the government. The only notable event is that *Artificial Intelligence* has received increased interest and that the French defense technology agency (Direction Générale de l'Armement—DGA) has decided to launch a call for projects on cybersecurity and artificial intelligence with a budget of 30 million Euros.

Brief Description

In France, the law programming the defense strategy for 2014–2019 is the main framework for the development of cybersecurity. It includes many of the aspects that need to be covered in the strategic research and innovation agenda. It creates obligations for operators of sensitive (critical) infrastructure to investigate and report incidents, and to be equipped to resist and/or detect attacks. It defines priority areas for the development of products that are either innovative or that include sovereignty issues. It also outlines a budget and describes several mechanisms (R&I funding, national platforms, strategic projects, ...) to reach these goals. This law can be seen as a model for the strategic development of cybersecurity in Europe [1].

The content of this section is focusing on more recent developments than were included in the EUNITY Grant Agreement, Annex 1 Description of the action [1]. The two fundamental documents included here are the draft military planning law (published in January 2018, for discussion by the French parliament) including the associated report and the defense review of October 2017, and the cyberdefense review published in February 2018. These documents form a consistent set and the core of the French cybersecurity doctrine for the 7 years to come (Table A.2).

Table A.2 Basic information about the National Cybersecurity Strategy in France

Owner (responsible)	French prime minister
Level of public administration—acceptance/publication	Government/Parliament
Date of publication	January 2018 (draft)
Type of document	Strategy
English name	Military planning law
Language versions	French and English summary[a]
Status	In force
Scope (geographical or sectoral)	French
Scope (on the basis of merit)	All aspects related to cybersecurity
Types of stakeholders	Industry, Critical infrastructure operators

[a]https://www.defense.gouv.fr/content/download/523961/9053454/file/MPL%202019-2025%20-%20Synopsis%20(EN).pdf

Previous Strategies and Related Documents

The military planning law is a 5-years exercise, that supports the long term planning of defense funding. As such, it is much wider than cybersecurity, and includes global security in its scope. Each law is usually created after the development of a strategic vision, developed in a "gold" book.

The laws have successively covered the 1997–2002, 2003–2008 and 2009–2014 periods. The 2009–2014 introduces a significant change, as it becomes the vehicle chosen by the French government to implement the cybersecurity strategy and the NIS directive. It includes in its scope a significant research effort of 110 million Euros. The current law covers the 2014–2019 period. Note that the spending planned in the law is usually adjusted in the yearly French budget. While the law provides a direction, its commitment to spending is weaker and usually not fully covered.

The law described thereafter covers the 2019–2025 period. It includes the text of the law, as well as a report evaluating the execution of the current period.

Strategic Vision

The strategic vision is that the effort implemented by the 2019–2025 law aims at adapting the French Armed Forces to the needs of France and Europe in 2030. It dedicates almost 200 billion Euros for the years 2019–2023, reflecting the French presidency objective to increase military spending to 2% of the French GDP by 2025.

The vision can be summarized by the seven following points:

1. place a priority on protecting our information systems;
2. adopt an active stance of attack deterrence and coordinated response;
3. fully exercise our digital sovereignty;
4. provide an effective penal response to cybercrime;

5. promote a shared culture of information security;
6. help bring about a digital Europe that is safe and reliable;
7. act internationally in favour of a collective and controlled governance of cyberspace.

The Main Aim

The law is organized in four clusters, two short term and two long term. The first cluster aims at improving soldier's life and training. The second cluster aims at improving and renewing aging military equipment. The third pole aims at guaranteeing France and Europe autonomy. The fourth pole focuses on innovation for future challenges. Cybersecurity is addressed in the last two poles.

Specific Aims

The law asserts that cyberspace is rapidly evolving and is having a dual role. Attack attribution is extremely difficult. The damages created by an attack on the activity of nation states make cybersecurity a priority challenge in the law.

The law thus aims at offering improved operational capabilities to the French government in cyberspace, particularly focused on intelligence gathering and cyberdefense. The law supports the capability to operate and innovate in cyberspace, considering that this is a strategic industrial sector for France. Intelligence gathering must become actionable, particularly bringing in the capability for attack attribution with sufficient certainty. It should also enable the evaluation of cyberattack capabilities of potential adversaries, and of the necessary means to defend against these cyberattacks.

It also aims at protecting France, the French government, the French armed forces and the critical infrastructure operators, against cyberattacks. It must develop tools and techniques to improve the resilience of the digital society.

It aims at creating a fifth branch of the military, devoted to cyberspace, with significant operational capabilities. This includes a significant effort on cyberdefense and protection of military networks. This also includes training for professionals, both for operations and for adaptation to new techniques or new threats, such as artificial intelligence.

The law also aims at sustaining the European strategic digital autonomy. This aim should include strategic partnerships, that can leverage capabilities at the EU level to provide a cumulative effect.

The cyberdefense review aims at creating four operational chains related to cybersecurity:

- **The "Protection" chain.** Under the authority of the Prime Minister, ANSSI is tasked with ensuring national security in the event of a cyberattack. It is supported by the military.

- **The "Military Action" chain**. Under the authority of the French President (head of the armed forces), it can use active cyber-warfare for military operations.
- **The "Intelligence" chain**. Covers actions related to intelligence gathering in cyberspace.
- **The "Judicial Investigation" chain**. Under the authority of the French Government, it develops investigative frameworks and coordinate the action of police under the control of the judicial system.

Areas of Interest—Research and Innovation Context

The strategy is supported by the need for France to master digital technologies. The strongest research and innovation point in the French strategy is related to artificial intelligence and big data, seen both as an opportunity for defense and as a threat (tool offered to attackers to improve their capabilities).

Three key digital technologies are particularly highlighted:

- **Encryption of communications**. France has a strong background on cryptography. Of particular interest are the aspects related to cloud data protection.
- **Detection of cyberattacks**. Several research and development efforts have been carried out to develop sovereign sensors for detecting attacks.
- **Professional mobile radio**. This effort focuses on security for 5G networks.

Areas of Interest—Legal, Policy and Organizational Perspective

The law defines an organizational basis for the French national cybersecurity agency (ANSSI) and the military central command (CALID) to monitor and protect cyberspace. It clearly states that international regulation is still very far away. While it acknowledges that negotiations have been ongoing since 2004, it also acknowledges the failure of the 2016–2017 expert groups meeting. One point of particular interest is the extraterritoriality of data, which is crossing French and European borders at will, and calls for new theoretical foundations for protection. It expresses a concern that this data is currently within the reach of the US judicial system only.

ANSSI is particularly tasked with improving the certification framework, in order to improve product security. The certification schemes, however, must become more practical than they are currently, and more visible to end-users. The European efforts related to certification provide an opportunity for EU-wide certification of products and services. However, these efforts must take into account the past history and ensure a high level of cybersecurity requirements.

Areas of Interest—Financial Perspective

The fundamental ambition of France is to develop a strong industrial base for cyber-security. Three areas of progress are particularly highlighted:

1. Encourage French industry leaders to develop their product offerings for cyber-security in the civilian sector,
2. Promote the creation of ETIs by helping the best SMEs grow, this action being supported by national investment funds in the area of cyberdefense,
3. Encourage start-up creation in the area of cyberdefense, this action being supported by specific technology accelerators and incubators.

France also supports the emergence of a European cybersecurity industrial base, including the emergence of an EU cybersecurity market, protected from foreign investment if necessary.

It also supports the development of insurance schemes, to support the continuous improvement of the cybersecurity posture, and information sharing.

The military planning law requires that the cyberdefense forces should reach about 4000 personnel, to which one should add the 750 forces devoted to the protection of the military digital infrastructure, and the 400 personnel devoted to export support.

The funding level will be increased from 730 million Euros in 2018 to 1 billion Euros in 2022.

Fulfilled Objectives

See Table A.3.

Examples of Actions in Line with the Strategy

The budgetary effort devoted to the French LPM aims at representing 2% of the French yearly Gross Domestic Product (GDP) by 2025.

To materialize the recent effort on Artificial Intelligence for Cybersecurity, the French military procurement agency (DGA) is launching a call for projects on data-lakes for cybersecurity and artificial intelligence for cybersecurity. The current plan at the time of writing is to devote 30 million euros to the call for proposals.

France has also reorganized the industrial landscape. It has created the *Comité Stratégique de Filière Sécurité* in May 2019. The scope of the organization is wider, because it encompasses all activities related to security, including civil protection, critical infrastructures and cybersecurity. The goal of the CSF Sécurité is to promote the French industry and products, particularly through the funding of large scale demonstrators. While the list of demonstrators is still under discussion, one of the demonstrators should include cybersecurity.

Table A.3 Fulfilled objectives by French National Cybersecurity Strategy

Objective	Status	Remarks
Set the vision, scope, objectives and priorities	✓	–
Follow a national risk assessment approach	✓	White Book, 2013. Strategic review, 2018
Take stock of existing policies, regulations and capabilities	✓	
Develop a clear governance structure	✓	Ongoing effort
Identify and engage stakeholders	✓	Focus on developing the industrial ecosystem
Establish trusted information-sharing mechanisms	–	Information sharing happens at the French and EU-bilateral level, but is not formally engaged by law. It is encouraged through insurance mechanisms
Develop national cyber contingency plans	✓	The current plans are focusing on government operations continuity
Protect critical information infrastructure	✓	249 critical infrastructure operators have been identified and are closely monitored and supported by ANSSI
Organise cyber security exercises	✓	In particular, FR is participating in NATO exercises
Establish baseline security requirements (measures)	✓	
Establish incident reporting mechanisms	✓	PDIS requirements, supported by insurance mechanisms
Raise user awareness	✓	Awareness programme managed by ANSSI
Strengthen training and educational programmes	✓	Training certification programme managed by ANSSI
Establish an incident response capability	✓	PRIS requirements
Address cyber crime	✓	However the mechanisms have not been laid out yet
Engage in international cooperation	✓	
Establish a public-private partnership	–	Engagement with regional clusters and industry associations
Balance security with privacy and data protection	✓	Implementation of GDPR through CNIL, PDIS requirements
Institutionalise cooperation between public agencies	✓	
Foster R&D	✓	Through all channels mentioned earlier
Provide incentives for the private sector to invest in security measures	✓	Regulation and support for critical infrastructure

Table A.4 Basic information about the National Cybersecurity Strategy in Spain

Owner (responsible entity)	National security department, presidency of the government of Spain
Level of public administration—acceptance/ publication	Government
Date of publication	2013
Type of document	Strategy
English name	National Cyber Security Strategy
Language versions	Spanish and English[a]
Status	In force
Scope (geographical or sectoral)	Spain
Scope (on the basis of merit)	All aspects related to cybersecurity
Types of stakeholders	Industry and research (All)

[a]https://www.enisa.europa.eu/topics/national-cyber-security-strategies/ncss-map/strategies/the-national-security-strategy/@@download_version/b8ba6b709f8f484db00d28d51d3b0452/file_en

Spain

Brief Description

In Spain, there is already legislation that covers some security requirements to be mapped to risk levels, namely the classification of information and the handling of such information is covered by law 9/1968, Law 11/2007 and Royal Decree 3/2010. There is no legislation yet that requires mandatory reporting of cybersecurity incidents, such as the one proposed in EU NIS directive, but the National Cyber Security Strategy, adopted in 2013, states that enforced incident reporting is a line of action that Spanish government will pursue. The Spanish National Cyber Security Strategy consists of the following five chapters:

- Cyberspace and security
- Purpose and guiding principles of cyber security in Spain
- Objectives of cyber security
- Lines of action of National cyber security
- Cyber security in the National Security System (Table A.4).

Strategic Vision

The National Cyber Security Strategy provides a basis for developing the provisions of the National Security Strategy of cyberspace, with the following main principles:

- National leadership
- Coordination of efforts
- Shared responsibility, Proportionality, Rationality and Efficiency
- International cooperation.

The outlined principles underline the importance of a development plan of the current situation, emphasizing the constitutional values protection as a common element.

The Main Aim

To ensure that Spain uses Information and Telecommunications Systems securely, by strengthening prevention, defence, detection, analysis, investigation, recovery and response capabilities vis-à-vis cyberattacks.

Specific Aims

The strategy sets up six specific objectives for different stakeholders. These specific aims are the following:

1. For the Public Authorities: Ensure an important level of security and resilience for the Information and Telecommunications Systems.
2. For Companies and Critical Infrastructures: To foster security and resilience of the networks and information systems used by the business sector in general and by operators of critical infrastructures in particular.
3. In the Judicial and police field operations: Enhance prevention, detection, response, investigation and coordination capabilities vis-à-vis terrorist activities and crime in cyberspace.
4. In the field of sensitisation: Raise awareness of citizens, professionals, companies and Spanish Public Authorities about cyberspace risks.
5. In capacity building: to gain and maintain the knowledge, skills, experience and technological capabilities Spain needs to underpin all the cyber security objectives
6. International collaboration: To contribute to improving cybersecurity in the international sphere.

Areas of Interest—Research and Innovation Context

The National Strategy defines several actions that could be analysed as areas of interest in research and innovation. In the line of action number 6, the document showcases the importance of promoting the training of professionals, giving crucial impetus to the industrial development by strengthening the Research and development efforts in cybersecurity matters. In these terms, the government created the Network of excellence on cybersecurity R&D+I, that brings together organisations such as INCIBE, dependent on the Spanish Ministry for Energy, Tourism and Digital Agenda, and others from the cybersecurity research ecosystem. This network set the following objectives:

- Boost the cooperation of expert agents in cybersecurity
- Collect research resources

- Disseminate research results
- Promote the creation of projects
- Build incentives to attract cybersecurity experts.

As a result of this initiative, the National Cybersecurity Cluster was created aiming at implementing several initiatives for coordination and knowledge transference between the R&D+I ecosystem and cybersecurity industry.

Areas of Interest—Legal, Policy and Organizational Perspective

The plan defines several areas of interest regarding legal perspective:

- The focus on combating cyber terrorism and crime should have an effective legal and operational framework. In this context, the importance of incorporating legal framework solutions to problems that arise in connection with cybersecurity in order to establish the different types of criminal infringements is outlined.
- The action line 4 underlines the importance of the development of a framework for cybersecurity knowledge.

Areas of Interest—Financial Perspective

The National Cyber Security Strategy does not define any topic regarding financial environment or funding for research.

Fulfilled Objectives

See Table A.5.

Examples of Actions in Line with the Strategy

The National Strategy of Cybersecurity develops the National Security Strategy plan of 2017 in cybersecurity, considering the general objectives, scope and action plan established for fulfilling it.

The document defining the strategy is composed of five chapters. The first one, called "The cyberspace, beyond a global common space", contains a vision of the scope of cybersecurity, the advances done so far since the approval of the Strategy in 2013, the reasons that make the National Strategy of Cybersecurity 2019 necessary and, finally, the main characteristics behind its development.

The activities performed in the cyberspace are fundamental for the actual society. The technology and infrastructure that compose the cyberspace are strategic elements, transversal to all activities. Being a vulnerability of the cyberspace is one of the biggest risks for the development of a nation. This is why, security in cyberspace

Table A.5 Fulfilled objectives by Spanish National Cybersecurity Strategy

Objective	Status	Remarks
Set the vision, scope, objectives and priorities	✓	
Follow a national risk assessment approach	✓	
Take stock of existing policies, regulations and capabilities	✓	
Develop a clear governance structure	✓	
Identify and engage stakeholders	✓	
Establish trusted information-sharing mechanisms	✓	
Develop national cyber contingency plans	✓	
Protect critical information infrastructure	✓	
Organise cyber security exercises	✓	
Establish baseline security requirements (measures)	✓	
Establish incident reporting mechanisms	✓	
Raise user awareness	✓	
Strengthen training and educational programmes	✓	
Establish an incident response capability	✓	
Address cyber crime	✓	
Engage in international cooperation	✓	
Establish a public/private partnership	✓	
Balance security with privacy and data protection	✓	
Institutionalise cooperation between public agencies	✓	
Foster R&D	✓	
Provide incentives for the private sector to invest in security measures	–	

is a key objective in the agenda of the governments in order to guarantee the National Security and it is a task of the State for creating a digital society that holds trust as a primary characteristic.

The second chapter, called "The threats and challenges in the cyberspace" describes the main threats of the cyberspace derived from its status as a common global space, of the highly technified society and connectivity incurring high impacts from any attack. It classifies attacks and challenges in two categories: threats for assets that are part of the cyberspace and, on the other hand, the ones using cyberspace as a mean for performing malicious activities.

The third chapter, called "Purpose and goals for cybersecurity" describe how the main action tasks of the National Security Strategy 2017 (action unit, anticipation, efficiency and resilience) are applied to five specific objectives. It is developed in chapter Four, called "Action tasks and measurement", where seven action tasks and the way to evaluate them are defined. These lines of action aim to reinforce the capacity of systems against threats from the cyberspace, guarantee the security and resilience of the strategic assets of Spain, boost cybersecurity of citizens and organisations, strengthen the research and pursuit of cybercriminals in order to guarantee the security and protection of freedom and rights in the cyberspace, impulse Spanish cybersecurity industry and the generation and retention of talent, contribute to international security in cyberspace, promote an open, plural, secure and trusted cyberspace that can support the national interests of the country and develop a cybersecurity culture that can support the National Security Culture Plan.

The fifth chapter, called "Cybersecurity in the National System" defines the architecture of cybersecurity at State level. Under the direction of the Prime Minister of the Government, the structure is composed of three departments: National Security Council, National Council of Cybersecurity and Status Committee. These three departments support each other and the Government, in particular in case of a cybersecurity crisis or any activity that goes beyond the normal behaviour of the departments (e.g. cyberattack to a critical infrastructure, targeted and/or coordinated cyberattacks to key organizations either public or private, etc.).

This system is complemented with the Cybersecurity Permanent Committee, which facilitates the inter-ministry coordination at operational level, being the department that assists the Cybersecurity National Council about technical and operative assessment and threats to cybersecurity, the public authorities and national CSIRT and the Cybersecurity National Forum.

Greece

The work presented here is an extension of the one presented in D3.1 "Preliminary version of the cybersecurity research analysis report for the two regions". This section updates and extends the information of that document with the news and changes since the publication of the previous report.

Brief Description

With the recent advances in digitization, the tremendous increase in data, the big data analytics, the increasing use of IoT and sensors, we are walking into an era of new arising cyber-security threats. The National Cyber Security Strategy intends to build and establish a cyber-secure and resilient online interactive environment for the citizens, the public and the private sector, in alignment with international standards, rules and good practices. The National Cyber Security Authority monitors, implements and handles the overall responsibility of the strategy along with National Advisory Body/ Forum. The National Computer Emergency Response Team (CERT) in cooperation with other entities, handles the operational level by dealing with the cyber attacks (Table A.6).

Table A.6 Basic information about the National Cybersecurity Strategy in Greece

Owner (responsible entity)	Ministry of digital policy, telecommunications and media
Level of public administration—acceptance/ publication	Council of Ministers (Government)
Date of publication	07/03/2018
Type of document	Strategy
English name	National Cyber Security Strategy
Language versions	Greek[a] and English[b]
Status	In force
Scope (geographical or sectoral)	Greece
Scope (on the basis of merit)	All aspects related to cybersecurity
Types of stakeholders	All

[a]https://www.enisa.europa.eu/topics/national-cyber-security-strategies/ncss-map/NCSSGR.pdf
[b]https://www.enisa.europa.eu/topics/national-cyber-security-strategies/ncss-map/GRNCSS_EN.pdf

Strategic Vision

This paragraph describes the national strategy concerning cybersecurity in Greece, setting the principles for the development of a safe online environment. The National Cyber Security Authority is implementing the National Cyber Security Strategy, that is called to fill the gap between the public and private sector. The National Cyber Security Strategy is basically a tool for assuring integrity, availability and resilience of critical frameworks, confidentiality of the transmitted information and subsequently advancement of online security.

The defence against cyber threats requires a set of principles and objectives in order to create a rule of law, ensuring high level of security and resilience, in alignment with principles like privacy, freedom, justice transparency and individual social rights.

The Main Aim

The aims of the National Cyber Security Strategy consist of the advancement of our protection mechanisms against cyber attacks, especially for the critical infrastructures, as well as the components to handle, recover from and minimize the impact. Alongside, it is important to keep an informed user space in private and public sector, aware of the cyber threats and protections mechanisms available. Specifically the strategy should initially address cyber crime, set a balance between security and privacy, provide critical information and infrastructure protection, enhance the development of the national cyber contingency plans, engage in international cooperation and establish a public-private partnership along with an incident response capability. The National Cyber Security Strategy should as well establish an institutionalised form of cooperation between public agencies and an implementation of policies and regulation capabilities, establish baseline security requirements, as well as the incident reporting mechanisms and trusted information-sharing mechanisms. Additionally, it should foster research and development, organise cyber security exercises, set a clear governance structure and strengthen the training and educational programmes. To accomplish these goals, Greece should keep up with international cyber security initiatives; measures, decisions and technological advancements of the international organizations and EU directions that guard the cyber space.

Specific Aims

- Increase effectiveness of national research systems: Enhancement of the level of prevention, assessment, study and recoverability of the threats against ICT systems including the strengthening of the public and private sectors with those capabilities
- Enrich the job research market
- Equal gender opportunity market in research
- Dissemination of scientific knowledge (open access in data and publications), including at national level the social institutions
- Make use of international practices
- Optimally initiate and increase interaction, cooperation and competitiveness with countries around the globe towards the upgrade of national security.

Areas of Interest—Research and Innovation Context

The actions that need to be undertaken include: the determination of the stakeholders that are part of the National Cyber Security Strategy in private and public sectors; the specification of the critical infrastructure in public and private sectors; the arrangement of the risk assessment study at national level including the identification, analysis and assessment of the impact of the risk. This will give the determination of the steps that can guard the critical infrastructure per sector/stakeholder that will be revised every three years. It should take in consideration risks associated with natural phenomena, technical problems or human error.

The fourth step includes the recording and enhancement of the existent institutional structures in order to be in alignment with the goals of the National Strategy. These include: the legislation, roles and capabilities of the stakeholders associated with Cyber Security as well as the administrative actions per sector; the structures, stakeholders and services in private and public sectors; the current emergency plans (Egnatia, Xenokrates etc.); and the European and international regulations on network and information security. Concerning the legislation enhancements, they should also cover the gaps in the constitutional freedoms and individual rights towards international Law.

Areas of Interest—Legal, Policy and Organizational Perspective

After the first ERA (European Research Area) monitoring report of European Commision in 2013, each Member State (MS) was obliged to prepare a national ERA for 2015–2020, which would define the financial research support of each country. The Greek government supported this decision and assigned to the Ministry of Education, Research and Religious affairs (MERRA) the task of capturing the current state of play, according to European objectives, setting future objectives (for 2020 and 2025) such as open access, building principles and policy directions.

A key instrument towards the ERA implementation is the Research and Innovation Strategy for Smart Specialisation 2014–2020 (RIS3). RIS3 includes the ERA principles and guidelines, the individual MS's strategic agendas as well as the Horizon 2020 thematic priorities along with its targets for joint actions and synergy. Of course ERA is not limited to cybersecurity and privacy.

Areas of Interest—Financial Perspective

Institutional Funding (public research organisations)—Project funding (jointly implemented by public and private bodies) State budget supports the permanent staff and other operating costs, as well as the payroll costs and operating expenses of public research institutions. They are calculated mainly by quantitative characteristics (non-competitive basis). On the other hand the public research organizations that constitute private law bodies and technological institutions are supported by the ordinary budget for their payroll expenses under the current legal framework. Here the programmatic agreements partly pay the operating costs.

In order to strengthen the funding structure described above there are several grants like the NSRF (National Strategic Reference Framework), which provides funds under specific actions ("Development proposals of Public Research Centers/institutions—KRIPIS"), EU's RTD Framework Programmes or other international programmes.

Fulfilled Objectives

See Table A.7.

Evaluation and Revision of the National Strategy

In order for the National Cyber Security Strategy to meet its goals, the National Cyber Security Authority should lead the evaluation and revision steps that will improve the Strategy. This evaluation will take into account the qualitative and quantitative efficiency alignment with international standards and practices within predefined time schedule.

Table A.7 Fulfilled objectives by Greek National Cybersecurity Strategy

Objective	Status	Remarks
Set the vision, scope, objectives and priorities	✓	
Follow a national risk assessment approach	✓	
Take stock of existing policies, regulations and capabilities	✓	
Develop a clear governance structure	✓	
Identify and engage stakeholders	✓	
Establish trusted information-sharing mechanisms	✓	In order to implement the National Strategy, an important aspect is the exchange of information between the stakeholders, concerning their security threats and policies applied. This measure should be applied both in private and public sector stakeholders, in order to analyse the advancements in national cyber security, alongside with the development of partnerships in public and private sectors, so a common ground of goals can be established
Develop national cyber contingency plans	✓	In case of important incidents in the critical information and the communication infrastructure of the stakeholders of the National Cyber Security Strategy, the national cyberspace contingency plan should define the arrangements, roles and competences of various stakeholders to deal with them
Protect critical information infrastructure	✓	
Organise cyber security exercises	✓	The simulation of security incidents is an important mechanism for the development of a defense mechanism, and drafted contingency plans as well as, the dissemination of information and knowledge with the participating stakeholders. These national preparedness exercises take place at regular intervals, are administered by the National Cyber Security Authority and their plans are resolved on specific timetables roles, scenarios and goals

(continued)

Table A.7 (continued)

Objective	Status	Remarks
Establish baseline security requirements (measures)	✓	All the stakeholders of the National Cyber Security Strategy should define the minimum security, technical and organizational measures, in order to establish an accepted level of security. Alongside, a common ground between the stakeholders should be established for the exchange of information, concerning the security incidents and measures to encounter them
Establish incident reporting mechanisms	✓	
Raise user awareness	✓	A very important aspect of ensuring cyber security in Greece, is the awareness and enhancing the knowledge about security threats, of the users of stakeholders and citizens in general. The National Cyber Security Authority is responsible for the planning and supervision of the user-citizen awareness programme, which includes citizen campaigns (in primary and secondary education via the Ministry of Education) educational actions for users and administrators (cooperation with university institutes) and promotion via websites
Strengthen training and educational programmes	✓	
Establish an incident response capability	✓	Concerning the security incident, the stakeholders are obliged to react effectively, apply the necessary actions, in order to prevent the repetition of the event, and spread this information, so the rest of the participant can be prepared. The administration of these episodes should follow the National Cyberspace Contingency Plan, and special teams are responsible for this like: CSIRTs (Computer Security Incident response Teams), in collaboration with the national CERT and other CSIRTs on national and international level, who should as well, optimize the level of prevention, evaluation and analysis of the threats
Address cyber crime	✓	

(continued)

Table A.7 (continued)

Objective	Status	Remarks
Engage in international cooperation	✓	Concerning cyber security, Greece should be effective in terms of international collaboration in order to keep up with the global directions and strategies, exchange experiences and best practices, especially with countries with similar planning. In order to enhance the cyber security at national level and align with international multilateral negotiations and actions, Greece should be active in the contribution and establishment of relevant agreements with international organizations
Establish a public/private partnership	✓	
Balance security with privacy and data protection	✓	
Institutionalise cooperation between public agencies	✓	
Foster R&D	✓	Support of research and development programmes and academic educational programmes. Taking in consideration that cyber security is a constantly evolving cutting edge technology, it is very important to keep a high level of support to the academic community in terms of research and development
Provide incentives for the private sector to invest in security measures	–	

Examples of Actions in Line with the Strategy

The development and the implementation phases consistute the main parts of the National Cyber Security Strategy, which are connected in a continuous cycle of implementation with two main levels of stakeholders involved, the strategic and the operational. The stakeholder at the higher political level implements the responsibility of the strategy and was established at the General Secretariat for Digital Policy of the Ministry of Digital Policy Telecommunications and Media. The National Cyber Security Authority along with the Computer Emergency Response Team (CERT) execute the tasks, where the public and the private sectors may also contribute. This scheme will monitor, coordinate and validate the stakeholders part. The actions required for the implementation of the strategy include the determination of the stakeholders that take part in the National Cyber Security Strategy, the definition of the critical infrastructure, the risk assessment following a scientific and technological procedure, the recording and improvement of the existing institutional framework, National Cyberspace Contingency Plan that will handle the significant incidents, the determination of the basic security requirements, the national preparedness exercises for locating the weaknesses and vulnerabilities of the system, the citizen awareness program, a trustworthy information exchange mechanism, the support of R&D programmes and academic educational programmes, the cooperation at international level and the evaluation and revision of the National Strategy.

Poland

Brief Description

The National Cybersecurity Strategy has been a subject of heated debate for a few years, as aspects of security (monitoring of traffic) need to be balanced with proper privacy controls. The draft document has been published for public debate by the Ministry of Digital Affairs in February 2016, but after revisions it has been approved in May 2017 as National Framework of Cybersecurity Policy of the Republic of Poland for 2017–2022.

The basic information of this document is listed in the following table and outlined in the following paragraphs (Table A.8).

Table A.8 Basic information about the National Cybersecurity Strategy in Poland

Owner (responsible entity)	Ministry of digital affairs
Level of public administration—acceptance/ publication	Council of Ministers (Government)
Date of publication	9 May 2017
Type of document	Strategy
English name	National Framework of Cybersecurity Policy of the Republic of Poland for 2017–2022
Language versions	Polish[a] and English[b]
Status	In force (Accepted by Polish Government)
Scope (geographical or sectoral)	Poland
Scope (on the basis of merit)	All aspects related to cybersecurity
Types of stakeholders	All: from the formal perspective—only public administration, but the document affects in some way all types of stakeholders (public administration, enterprises, individuals) in Poland (also foreign enterprises which operate in Poland)

[a]https://www.gov.pl/documents/31305/0/krajowe_ramy_polityki_cyberbezpieczenstwa_rzeczypospolitej_polskiej_na_lata_2017_-_2022.pdf
[b]https://www.enisa.europa.eu/topics/national-cyber-security-strategies/ncss-map/Cybersecuritystrategy_PL.pdf

Strategic Vision

In 2022 Poland will be more resilient from attacks and threats from cyberspace. Due to synergy in internal and international actions, the cyberspace of the Republic of Poland will be a safe environment to make use of digital economy with respect to rights and freedoms of citizens.

The Main Aim

Ensuring high level of security of the public sector, private sectors and citizens in the area of essential services and digital services.

Specific Aims

Four detailed aims are provided, such as:

1. Gaining the ability of coordinated actions at the national level, which aims at prevention, detection, fighting and minimization of impacts of security incidents.
2. Strengthening the ability of cyber threats prevention.
3. Enhancement of national potential and competences in the area of security in cyberspace
4. Building of strong international position of Poland in the area of cybersecurity.

Areas of Interest—Research and Innovation Context

In the strategy, actions are defined which aim at investing in the development of industrial and technological resources for cybersecurity, by creating the conditions needed for the development of enterprises, scientific research centres and start-ups in the area of cybersecurity. The creation of the Cyberpark Enigma programme, which aims at producing high quality hardware and software is provided as an example of such actions.

Another action, which is planned in order to increase the competence of research centres in the area of cybersecurity, is the creation of a Scientific Cybersecurity Cluster (SCC), which will be a scientific platform composed of higher education institutions and scientific research centres specialising in cybersecurity technologies.

The creation of innovation hubs, which will offer comprehensive services for companies and start-ups, including testing new solutions, market research, support in applying for funding for development of innovation solutions, advice on access to new markets and assistance in establishing cooperation with other entrepreneurs is also planned.

The areas of research in the field of cybersecurity which were identified in the strategy are the following:

- transformation from IPv4 currently used on the Internet to IPv6,
- the Internet of Things,
- Smart Cities,
- Industry 4.0,
- Cloud Computing,
- Big Data.

Specific research programmes will be also prepared in order to:

- prepare and implement new methods of protection against novel threats from cyberspace,
- assess the effectiveness of protections and resistance of Poland's cyberspace to cyberthreats,
- assess the effectiveness of response to threats,
- analyse new trends in cybercrime, cyberterrorism and methods of combating them,
- study methods of attacks and ways of counteracting these attacks.

Areas of Interest—Legal, Policy and Organizational Perspective

The main areas of interest, described by the document, are the following:

- The need for legal changes, changes in the structure of cybersecurity system and improvements in effectiveness of interactions of institutions responsible for cybersecurity to outline clear areas of responsibility.
- The need for cybersecurity risk management at the national level. The need of critical infrastructure, essential services and digital services protection, as well as cybercrime and cyberterrorism prevention.
- The need for the ability to perform military actions in cyberspace.
- The need to build competences in the area of cybersecurity: building public-private partnerships, stimulation of research and development and building awareness in the area of cybersecurity.

The document puts cybersecurity on equal level to the military security of the nation.

Areas of Interest—Financial Perspective

In the strategy there is no specific information regarding funding, however the need of preparing a research programme aimed at preparation and implementation of new methods of protection against novel threats from cyberspace, which will be launched jointly with the National Centre for Research and Development, is indicated. Also other research programmes are to be developed with the scientific and academic community.

Fulfilled Objectives

See Table A.9.

Next Steps

The document provides only general aims, guidelines and declarations. Because of that current works are concentrated on preparing an Action Plan, which will implement these aims. After two and after four years from publishing this act there is a plan to conduct reviews and assessment of the effects of this act. The results will be presented to the Council of Ministers, and the Minister of Digital Affairs will be obliged to prepare propositions of corrective actions or a draft of a document which will be in force during the next 5 years.

Examples of Actions in Line with the Strategy

The Polish Parliament passed the Act on the National Cybersecurity System, which has entered into force on 28 August 2018. The Act implements the provisions of the NIS Directive and introduces a national cybersecurity system. Several research projects conducted in Poland in the area of cybersecurity and privacy are in line with the strategy and also with the Act. For example, the National Cybersecurity Platform is a very big research and development project, which aims to create an interactive system of monitoring and visualization of the actual security status of national cyberspace. The system will use various methods of dynamic and static risk analyses, expert system for decision support as well as many tools for vulnerabilities and threats detection.

Belgium

Brief Description

Belgium's Cyber Security Strategy was adopted by the government in 2012. The strategy document, prepared by BelNIS (Belgian Network and Information Security; a national consultation forum on information security), identifies three strategic objectives which are to be realised through different initiatives in eight identified action domains. These three strategic objectives are (1) a safe and reliable cyberspace, (2) an optimal security and protection for critical infrastructures and governmental information systems, and (3) the development of national cybersecurity capabilities. The legal framework for cybersecurity in Belgium, however, remains somewhat unclear, and the information available on the implementation of the strategy is limited. Belgium also does not have a comprehensive critical infrastructure protection

Table A.9 Fulfilled objectives by Polish National Cybersecurity Strategy

Objective	Status	Remarks
Set the vision, scope, objectives and priorities	✓	The strategic vision as well as the main goal and four specific objectives are set, however the strategy itself does not provide prioritization of the aims
Follow a national risk assessment approach	✓	Development and implementation of a risk management system at the national level is one of the defined actions in the first specific objective. The document states that a coherent risk assessment methodology, taking into account the specificity of individual sectors and operators of critical infrastructure, key services and digital service providers, will be developed
Take stock of existing policies, regulations and capabilities	✓	In the strategy only few of existing regulations (for example the NIS Directive) are taken into account. But one of the main purposes of the first specific objective defined in the strategy is conducting a review of sectoral and specific regulations in the area of cybersecurity
Develop a clear governance structure	✓	Improving the structure of the national cybersecurity system is one of the actions defined by the first specific objective
Identify and engage stakeholders	✓/–	In the strategy the need for identification and engaging stakeholders is not literally identified, however this objective could be satisfied on the basis of other defined actions (such as: increasing the effectiveness of cooperation of entities ensuring security)
Establish trusted information-sharing mechanisms	✓	The need for a system of cyberspace security management at the national level, which will allow threats and vulnerabilities to be reported, is identified in the strategy. The need to set up systemic solutions to exchange information between stakeholders and share knowledge about threats and incidents is also identified
Develop national cyber contingency plans	–	This objective is defined only indirectly
Protect critical information infrastructure	✓	The document states that the government will take action to improve the capacity and competence in cybersecurity of critical infrastructure operators

(continued)

Table A.9 (continued)

Objective	Status	Remarks
Organise cyber security exercises	✓	The need for many types of exercises is indicated, such as: comprehensive exercises simulating the nationwide incidents, smaller-scale exercises, including sectoral ones, exercises of the Armed Forces of the Republic of Poland to carry out military operations in cyberspace and also international exercises
Establish baseline security requirements (measures)	✓	One of actions indicated in the document regards the development of minimum requirements for ICT security, which will also cover the business continuity management
Establish incident reporting mechanisms	✓	One of the main aims defined in the document is the development of the cybersecurity system at the national level, which also entails the further development of structures dealing with cybersecurity at the operational level, including CSIRT teams at the national level, sectoral incident response teams (sectoral CSIRT), information exchange and analysis centres (ISAC)
Raise user awareness	✓	Many objectives in the area of raising user awareness are defined. Users in this context are perceived as IT staff of companies, IT staff of public administration as well as ordinary citizens
Strengthen training and educational programmes	✓	The need for many actions towards training and education is identified in the strategy
Establish an incident response capability	✓	An incident response capability is to be established on the basis of CSIRT teams.
Address cyber crime	✓	The strategy defines a set of actions to enhance the capacity to counteract cybercrime, including cyberespionage, incidents of a terrorist nature and hybrid threats
Engage in international cooperation	✓	Many types of cooperation are listed in the document: at the European level, in the NATO and within United Nations. The cooperation also at the operational and technical level is identified as an important action to carry out
Establish a public/private partnership	✓	The goal of building cooperation mechanisms between the public sector and the private sector is defined

(continued)

Table A.9 (continued)

Objective	Status	Remarks
Balance security with privacy and data protection	–	The objective is defined only indirectly
Institutionalise cooperation between public agencies	✓	The objective is met by various proposed actions
Foster R&D	✓	Defined above
Provide incentives for the private sector to invest in security measures	✓/–	This objective is partially met by specific actions (such as creation innovation hubs)

Table A.10 Basic information about the National Cybersecurity Strategy in Belgium

Owner (responsible entity)	Centre for cybersecurity (Belgium)
Level of public administration—acceptance/publication	Government
Date of publication	2012
Type of document	Strategy
English name	Cyber Security Strategy—Securing Cyberspace
Language versions	French[a]
Status	In force
Scope (geographical or sectoral)	Across the whole of Belgian territories
Scope (on the basis of merit)	All aspects related to cybersecurity
Types of stakeholders	All

[a]https://www.enisa.europa.eu/topics/national-cyber-security-strategies/ncss-map/ncss-be-fr

strategy or plan in place, but Sect. 3 of the Cyber Security Strategy addresses, in part, the need for the government to work with entities engaged with critical infrastructure for the purposes of data protection and the development of improved incident management procedures.

On the other hand, Belgium does have an established computer emergency response team, CERT.be, and a well-developed cybersecurity incident-reporting structure. In the October 2015, Belgium's federal government officially launched the Belgian Centre for Cyber Security. One of the organisation's main missions will be to further educate all individuals and organisations in the importance of cyber security. Another aim of the organisation is to streamline and co-ordinate the various government and/or private initiatives around cyber security.

There is active support in the country for public-private partnerships, initially through BeINIS, nowadays through the aforementioned Centre for Cyber Security, on the one hand, and the Belgian Cyber Security Coalition (a partnership between players from the academic world, the public authorities and the private sector to join forces in the fight against cybercrime), on the other hand [4] (Table A.10).

Strategic Vision

To identify three strategic objectives which are to be realised through different initiatives in eight identified action domains. These three strategic objectives are (1) a safe and reliable cyberspace, (2) an optimal security and protection for critical infrastructures and governmental information systems, and (3) the development of national cybersecurity capabilities [4].

The Main Aim

To ensure and secure the cyberspace for Belgian players.

Specific Aims

Specific aims are hereby grouped: cyber threats, strategic objectives, approach and domains of action.

Areas of Interest—Research and Innovation Context

Stimulation of the development of new technologies via access to major hi-tech projects, fostering collaborative efforts in a multi stakeholder research agenda.

Areas of Interest—Legal, Policy and Organizational Perspective

The legal framework for cybersecurity in Belgium, remains somewhat unclear, and the information available on the implementation of the strategy is limited [1].

Areas of Interest—Financial Perspective

None to record.

Fulfilled Objectives

See Table A.11.

Table A.11 Fulfilled objectives by Belgian National Cybersecurity Strategy

Objective	Status	Remarks
Set the vision, scope, objectives and priorities	✓	Vision, goal and clear objectives are all well defined
Follow a national risk assessment approach	✓/–	Only indirectly mentioned
Take stock of existing policies, regulations and capabilities	–	No mention
Develop a clear governance structure	–	No mention
Identify and engage stakeholders	✓	The strategy aims at reaching out to all stakeholders and engage with them
Establish trusted information-sharing mechanisms	–	No mention
Develop national cyber contingency plans	✓	Indirectly set forth
Protect critical information infrastructure	✓	Annex on critical infrastructures included
Organise cyber security exercises	–	No direct mention
Establish baseline security requirements (measures)	–	Establishment of guidance on policy settings rather than baseline security requirements
Establish incident reporting mechanisms	–	No direct mention
Raise user awareness	✓	Strategic goal to achieve
Strengthen training and educational programmes	✓	Raising awareness is within the main objectives
Establish an incident response capability	–	No direct mention
Address cyber crime	✓	Substantive section in the strategy
Engage in international cooperation	✓/–	Indirectly mentioned
Establish a public/private partnership	–	Not directly from the strategy
Balance security with privacy and data protection	–	No mention
Institutionalise cooperation between public agencies	✓/–	Indirect
Foster R&D	✓	Substantive section in the strategy
Provide incentives for the private sector to invest in security measures	–	No direct mention

Table A.12 Basic information about the National Cybersecurity Strategy in Austria

Level of public administration—acceptance/ publication	Federal chancellery of the Republic of Austria (Government)
Date of publication	March 2013
Type of document	Strategy
English name	Austrian Cyber Security Strategy
Language versions	Austrian[a] and English[b]
Status	In force
Types of stakeholders	All

[a]https://www.digitales.oesterreich.gv.at/at.gv.bka.liferay-app/documents/22124/30428/Oesterre ichischeStrategieCyber-Sicherheit.pdf/fd94cf23-719b-4ef1-bf75-385080ab2440
[b]https://www.digitales.oesterreich.gv.at/at.gv.bka.liferay-app/documents/22124/30428/Austrian CyberSecurityStrategy.pdf/35f1c891-ca99-4185-9c8b-422cae8c8f21

Next Steps

The strategy does not provide substantive nor granular explanations on the roll-out of the strategy. The policy is quite short and high-level in the objectives is intended to set forth.

Action Plan

The implementation of the NIS Directive, particularly with regard to critical infras-tructures, will be a crucial functional part in the achievement of the strategy. CERT was established and now the next phases will be to enhance intra-national coopera-tion with private business as well as maximize the collaboration with CERTs from other Member States of the Union.

Other Countries

Austria

Austrian Cyber Security Strategy is the main strategic document in the area of cybersecurity. The basic information of this document is listed in the following (Table A.12).

Bulgaria

National Cyber Security Strategy "Cyber Resilient Bulgaria 2020" is the main strate-gic document in the area of cybersecurity.

Table A.13 Basic information about the National Cybersecurity Strategy in Bulgaria

Level of public administration—acceptance/publication	Council of Ministers
Date of publication	18 July 2016
Type of document	Strategy
English name	National Cyber Security Strategy "Cyber Resilient Bulgaria 2020"
Language versions	Bulgarian[a]
Status	In force

[a]https://www.enisa.europa.eu/topics/national-cyber-security-strategies/ncss-map/strategies/national-cyber-security-strategy-6/@@download_version/5024530da77b4058bf2d7d0389c5a2a8/file_en

Table A.14 Basic information about the National Cybersecurity Strategy in Croatia

Level of public administration—acceptance/publication	(Government)
Date of publication	7 October 2015
Type of document	Strategy
English name	The National Cyber Security Strategy of the Republic of Croatia
Language versions	English[a]
Status	In force

[a]https://www.enisa.europa.eu/topics/national-cyber-security-strategies/ncss-map/CRNCSSEN.pdf

The basic information of this document is listed in the following (Table A.13).

Croatia

The National Cyber Security Strategy of the Republic of Croatia is the main strategic document in the area of cybersecurity.

The basic information of this document is listed in the following (Table A.14).

Cyprus

Cybersecurity Strategy of the Republic of Cyprus is the main strategic document in the area of cybersecurity. The basic information of this document is listed in the following (Table A.15).

Table A.15 Basic information about the National Cybersecurity Strategy in Cyprus

Level of public administration—acceptance/ publication	(Government)
Date of publication	23 April 2012
Type of document	Strategy
English name	Cybersecurity Strategy of the Republic of Cyprus: Network and Information Security and Protection of Critical Information Infrastructures
Language versions	English[a]
Status	In force

[a]https://www.enisa.europa.eu/topics/national-cyber-security-strategies/ncss-map/CybersecuritySt rategyoftheRepublicofCyprusv10_English.pdf

Table A.16 Basic information about the National Cybersecurity Strategy in the Czech Republic

Level of public administration—acceptance/ publication	(Government)
Date of publication	1 January 2015
Type of document	Strategy
English name	National Cyber Security Strategy of the Czech Republic for the period from 2015 to 2020
Language versions	English[a] and Czech[b]
Status	In force

[a]https://www.enisa.europa.eu/topics/national-cyber-security-strategies/ncss-map/CzechRepublic _Cyber_Security_Strategy.pdf
[a]https://www.enisa.europa.eu/topics/national-cyber-security-strategies/ ncss-map/strategies/cyber-security-strategy-of-czech-republic-2011-2015/@@download_versi on/48c136b4728d4a05aad610a436719ae0/file_native

Czech Republic

National Cyber Security Strategy of the Czech Republic for the period from 2015 to 2020 is the main strategic document in the area of cybersecurity. The basic information of this document is listed in the following (Table A.16).

Denmark

The basic information of this document is listed in the following (Table A.17).

Estonia

The basic information of the strategy is listed in the following (Table A.18).

Table A.17 Basic information about the National Cybersecurity Strategy in Denmark

Level of public administration—acceptance/publication	Ministry of finance
Date of publication	May 2018
Type of document	Strategy
English name	The Danish Cyber and Information Security Strategy 2018–2021
Language versions	English[a] and Danish[b]
Status	In force

[a]https://www.enisa.europa.eu/topics/national-cyber-security-strategies/ncss-map/strategies/natio
nal-strategy-for-cyber-and-information-security/@@download_version/8b31862c3e304fceadc
0719d18dc3bb3/file_en
[a]https://www.enisa.europa.eu/topics/national-cyber-security-strategies/ncss-map/strategies/natio
nal-strategy-for-cyber-and-information-security/@@download_version/8b31862c3e304fce
adc0719d18dc3bb3/file_native

Table A.18 Basic information about the National Cybersecurity Strategy in Estonia

Level of public administration—acceptance/publication	Ministry of economic affairs and communication
Date of publication	2014
Type of document	Strategy
English name	Cyber Security Strategy
Language versions	English[a] and Estonian[b]
Status	In force

[a]https://www.enisa.europa.eu/topics/national-cyber-security-strategies/ncss-map/strategies/cy
ber-security-strategy/@@download_version/ee51977af0414064a5b35d0fa63e15f1/file_en
[a]https://www.enisa.europa.eu/topics/national-cyber-security-strategies/ncss-map/strategies/cyb
er-security-strategy/@@download_version/ee51977af0414064a5b35d0fa63e15f1/file_native

Table A.19 Basic information about the National Cybersecurity Strategy in Finland

Level of public administration—acceptance/publication	Security and defence committee—ministry of defence
Date of publication	24 January 2013
Type of document	Strategy
English name	Finland's Cyber security Strategy
Language versions	English[a]
Status	In force

[a]https://www.enisa.europa.eu/topics/national-cyber-security-strategies/ncss-map/strategies/fin
lands-cyber-security-strategy/@@download_version/ec1473d25aac4b6f8771ff0825834e05/file_en

Finland

The basic information of the strategy is listed in the following (Table A.19).

Table A.20 Basic information about the National Cybersecurity Strategy in Germany

Level of public administration—acceptance/ publication	(Government)
Date of publication	2011
Type of document	Strategy
English name	Cyber Security Strategy for Germany
Language versions	English[a] and German[b]
Status	In force

[a]https://www.enisa.europa.eu/topics/national-cyber-security-strategies/ncss-map/strategies/cyb er-security-strategy-for-germany/@@download_version/8adc42e23e194488b2981ce41d9de93e/ file_en
[b]https://www.enisa.europa.eu/topics/national-cyber-security-strategies/ncss-map/strategies/cyb er-security-strategy-for-germany/@@download_version/8adc42e23e194488b2981ce41d9de93e/ file_native

Table A.21 Basic information about the National Cybersecurity Strategy in Hungary

Level of public administration—acceptance/ publication	Government
Date of publication	21 March 2013
Type of document	Strategy
English name	National Cyber Security Strategy of Hungary
Language versions	English[a]
Status	In force

[a]https://www.enisa.europa.eu/topics/national-cyber-security-strategies/ncss-map/HU_NCSS.pdf

Germany

The basic information of the strategy is listed in the following (Table A.20).

Hungary

The basic information of the strategy is listed in the following (Table A.21).

Ireland

The basic information of the strategy is listed in the following (Table A.22).

Italy

The basic information of the strategy is listed in the following (Table A.23).

Table A.22 Basic information about the National Cybersecurity Strategy in Ireland

Level of public administration—acceptance/publication	Department of communications, energy and natural resources (Government)
Date of publication	Unknown
Type of document	Strategy
English name	National Cyber Security Strategy 2015–2017
Language versions	English[a]
Status	Unknown (outdated)

[a]https://www.enisa.europa.eu/topics/national-cyber-security-strategies/ncss-map/NCSS_IE.pdf

Table A.23 Basic information about the National Cybersecurity Strategy in Italy

Level of public administration—acceptance/publication	Presidency of the council of ministers (Government)
Date of publication	December 2013
Type of document	Strategy
English name	National Strategic Framework for Cyberspace Security
Language versions	English[a]
Status	In force

[a]https://www.enisa.europa.eu/topics/national-cyber-security-strategies/ncss-map/IT_NCSS.pdf
[b]https://www.enisa.europa.eu/topics/national-cyber-security-strategies/ncss-map/strategies/national-strategic-framework-for-cyberspace-security/@@download_version/2f91075f0f7f404eb1ae
bd465a553629/file_native

Table A.24 Basic information about the National Cybersecurity Strategy in Latvia

Level of public administration—acceptance/publication	(Government)
Date of publication	Unknown (implementation date: June 2014)
Type of document	Strategy
English name	Cyber Security Strategy of Latvia 2014–2018
Language versions	English[a] and Latvian[b]
Status	In force

[a]https://www.enisa.europa.eu/topics/national-cyber-security-strategies/ncss-map/lv-ncss
[b]https://www.enisa.europa.eu/topics/national-cyber-security-strategies/ncss-map/strategies/latvian-national-cyber-security-strategy/@@download_version/f05ac35addcb4f9f8bcb748f0e782cbc/
file_native

Latvia

The basic information of the strategy is listed in the following (Table A.24).

Table A.25 Basic information about the National Cybersecurity Strategy in Lithuania

Level of public administration—acceptance/ publication	Government
Date of publication	29 June 2011
Type of document	Strategy
English name	Programme for the development of electronic information security (cyber security) for 2011–2019
Language versions	English[a]
Status	In force

[a]https://www.enisa.europa.eu/topics/national-cyber-security-strategies/ncss-map/Lithuania_ Cyber_Security_Strategy.pdf

Table A.26 Basic information about the National Cybersecurity Strategy in Luxembourg

Level of public administration—acceptance/ publication	Government council
Date of publication	26 January 2018 (date of approval)
Type of document	Strategy
English name	National Cyber Security Strategy
Language versions	English[a] and French[b]
Status	In force

[a]https://www.enisa.europa.eu/topics/national-cyber-security-strategies/ncss-map/nationalcyber securitystrategyiiien.pdf
[b]https://www.enisa.europa.eu/topics/national-cyber-security-strategies/ncss-map/strategies/strate gie-nationale-en-matiere-de-cyber-securite/@@download_version/d4af182d7c6e4545ae751c17 fcca9cfe/file_native

Lithuania

The basic information of the strategy is listed in the following (Table A.25).

Luxembourg

The basic information of the strategy is listed in the following (Table A.26).

Malta

The basic information of the strategy is listed in the following (Table A.27).

Table A.27 Basic information about the National Cybersecurity Strategy in Malta

Level of public administration—acceptance/publication	Government
Date of publication	Unknown
Type of document	Strategy
English name	Malta Cyber Securty Strategy
Language versions	English[a]
Status	In force

[a]https://www.enisa.europa.eu/topics/national-cyber-security-strategies/ncss-map/Mita_MaltaCyberSecurityStrategy.pdf

Table A.28 Basic information about the National Cybersecurity Strategy in the Netherlands

Level of public administration—acceptance/publication	Minister of justice and security (Government)
Date of publication	Unknown (2018)
Type of document	Strategy
English name	National Cyber Security Agenda
Language versions	English[a] and Dutch[b]
Status	In force

[a]https://www.enisa.europa.eu/topics/national-cyber-security-strategies/ncss-map/CSAgenda_EN_nl.pdf
[b]https://www.enisa.europa.eu/topics/national-cyber-security-strategies/ncss-map/strategies/national-cyber-security-strategy-1/@@download_version/82b3c1a34de449f48cef8534b513caea/file_native

Table A.29 Basic information about the National Cybersecurity Strategy in Portugal

Level of public administration—acceptance/publication	Government
Date of publication	28 May 2015
Type of document	Strategy
English name	National Cyberspace Security Strategy
Language versions	English[a]
Status	In force

[a]https://www.enisa.europa.eu/topics/national-cyber-security-strategies/ncss-map/Portuguese_National_Cyberspace_Security_Strategy_EN.pdf

The Netherlands

The basic information of the strategy is listed in the following (Table A.28).

Portugal

The basic information of the strategy is listed in the following (Table A.29).

Table A.30 Basic information about the NCSS in Romania

Level of public administration—acceptance/ publication	Government
Date of publication	23 May 2013
Type of document	Strategy
English name	- (Cyber Security Strategy in Romania)
Language versions	English[a] and Romanian[b]
Status	In force

[a]https://www.enisa.europa.eu/topics/national-cyber-security-strategies/ncss-map/strategies/cyber-security-strategy-in-romania/@@download_version/1b41c7f470b14b52be67866e84007f87/file_en
[b]https://www.enisa.europa.eu/topics/national-cyber-security-strategies/ncss-map/StrategiaDeSecuritateCiberneticaARomaniei.pdf

Table A.31 Basic information about the NCSS in Slovakia

Level of public administration—acceptance/ publication	Government
Date of publication	Unknown
Type of document	Strategy
English name	Cyber Security Concept of the Slovak Republic and Action Plan of Implementation of Cyber Security Concept of the Slovak Republic
Language versions	English[a] and Slovak[b]
Status	In force

[a]https://www.enisa.europa.eu/topics/national-cyber-security-strategies/ncss-map/cyber-security-concept-of-the-slovak-republic-1
https://www.enisa.europa.eu/topics/national-cyber-security-strategies/ncss-map/ActionPlanfortheImplementationoftheCyberSecurityConceptoftheSlovakRepublicfor20152020_3_.pdf
[b]https://www.enisa.europa.eu/topics/national-cyber-security-strategies/ncss-map/strategies/cyber-security-concept-of-the-slovak-republic/@@download_version/92bd73c35c5d47a9ac1d8d8dc282d318/file_native

Romania

The basic information of the strategy is listed in the following (Table A.30).

Slovakia

The basic information of the strategy is listed in the following (Table A.31).

Slovenia

The basic information of the strategy is listed in the following (Table A.32).

Table A.32 Basic information about the National Cybersecurity Strategy in Slovenia

Level of public administration—acceptance/ publication	Government
Date of publication	February 2016
Type of document	Strategy
English name	Cyber Security Strategy
Language versions	English[a] and Slovene[b]
Status	In force

[a]https://www.enisa.europa.eu/topics/national-cyber-security-strategies/ncss-map/si-ncss
[b]https://www.enisa.europa.eu/topics/national-cyber-security-strategies/ncss-map/DSI2020_Strategija_Kibernetske_Varnosti.pdf

Table A.33 Basic information about the National Cybersecurity Strategy in Sweden

Level of public administration—acceptance/ publication	Government
Date of publication	22 June 2016
Type of document	Strategy
English name	National Cyber Security Strategy
Language versions	English[a] and Swedish[b]
Status	In force

[a]https://www.enisa.europa.eu/topics/national-cyber-security-strategies/ncss-map/SwedishNCSSen.pdf
[b]https://www.enisa.europa.eu/topics/national-cyber-security-strategies/ncss-map/swedish-ncss

Table A.34 Basic information about the National Cybersecurity Strategy in the United Kingdom

Level of public administration—acceptance/ publication	Government
Date of publication	2016
Type of document	Strategy
English name	National Cyber Security Strategy 2016–2021
Language versions	English[a]
Status	In force

[a]https://www.enisa.europa.eu/topics/national-cyber-security-strategies/ncss-map/national_cyber_security_strategy_2016.pdf

Sweden

The basic information of the strategy is listed in the following (Table A.33).

United Kingdom

The basic information of the strategy is listed in the following (Table A.34).

Fulfilled Objectives

Table A.35 presents fullfilled objectives by the EU Member States (legend provided in Table A.36).

Table A.35 Fulfilled objectives by NCSSs

Objective	AT	BG	HR	CY	CZ	DK	EE	FI	DE	HU	IE
Total number of objectives defined	12	8	7	10	9	11	13	14	9	9	5
Set the vision, scope, objectives and priorities	–	–	–	–	–	–	–	–	–	–	–
Follow a national risk assessment approach	–	–	–	–	–	–	–	–	–	–	–
Take stock of existing policies, regulations and capabilities	–	–	–	–	–	–	–	–	–	–	–
Develop a clear governance structure	–	–	–	–	–	–	–	–	–	–	–
Identify and engage stakeholders	–	–	–	–	–	–	–	–	–	–	–
Establish trusted information-sharing mechanisms	–	–	–	–	–	–	–	–	–	–	–
Develop national cyber contingency plans	✓	–	–	✓	–	✓	✓	✓	✓	✓	✓
Protect critical information infrastructure	✓	–	–	✓	✓	✓	✓	✓	✓	✓	–
Organise cyber security exercises	✓	–	–	✓	✓	✓	✓	✓	–	✓	✓
Establish baseline security requirements (measures)	✓	–	✓	–	–	✓	✓	✓	✓	✓	–
Establish incident reporting mechanisms	✓	✓	✓	–	–	✓	✓	✓	–	✓	✓
Raise user awareness	–	✓	–	✓	✓	✓	✓	✓	✓	–	–
Strengthen training and educational programmes	✓	✓	–	✓	–	✓	✓	✓	–	✓	–
Establish an incident response capability	✓	✓	✓	✓	✓	✓	✓	✓	✓	✓	✓
Address cyber crime	✓	✓	✓	✓	–	✓	✓	✓	✓	–	–
Engage in international cooperation	✓	✓	✓	✓	✓	✓	✓	✓	✓	✓	✓
Establish a public/private partnership	✓	–	–	✓	✓	–	–	✓	–	–	–
Balance security with privacy and data protection	–	✓	✓	–	✓	–	✓	✓	–	–	–
Institutionalise cooperation between public agencies	✓	–	–	✓	✓	–	✓	–	✓	✓	–
Foster R&D	✓	✓	✓	–	✓	✓	✓	✓	✓	–	–
Provide incentives for the private sector to invest in security measures	–	–	–	–	–	–	–	✓	–	–	–

(continued)

Table A.35 (continued)

Objective	IT	LV	LT	LU	MT	NL	PT	RO	SK	SI	SE	UK
Total number of objectives defined	15	9	8	12	6	7	11	11	11	10	9	12
Set the vision, scope, objectives and priorities	–	–	–	–	–	–	–	–	–	–	–	–
Follow a national risk assessment approach	–	–	–	–	–	–	–	–	–	–	–	–
Take stock of existing policies, regulations and capabilities	–	–	–	–	–	–	–	–	–	–	–	–
Develop a clear governance structure	–	–	–	–	–	–	–	–	–	–	–	–
Identify and engage stakeholders	–	–	–	–	–	–	–	–	–	–	–	–
Establish trusted information-sharing mechanisms	–	–	–	–	–	–	–	–	–	–	–	–
Develop national cyber contingency plans	✓	✓	✓	✓	–	–	–	✓	–	–	✓	–
Protect critical information infrastructure	✓	✓	✓	✓	–	✓	✓	✓	✓	✓	✓	✓
Organise cyber security exercises	✓	–	–	✓	–	–	✓	–	✓	✓	✓	✓
Establish baseline security requirements (measures)	✓	✓	✓	✓	–	✓	✓	✓	✓	–	–	✓
Establish incident reporting mechanisms	✓	–	–	✓	✓	–	✓	✓	✓	✓	–	✓
Raise user awareness	✓	✓	✓	✓	✓	–	✓	✓	✓	✓	–	✓
Strengthen training and educational programmes	✓	✓	✓	✓	✓	✓	✓	✓	✓	✓	–	✓
Establish an incident response capability	✓	✓	✓	✓	✓	✓	✓	✓	✓	✓	–	✓
Address cyber crime	✓	✓	✓	✓	✓	✓	✓	✓	–	✓	✓	✓
Engage in international cooperation	✓	✓	✓	✓	✓	✓	✓	✓	✓	✓	✓	✓
Establish a public/private partnership	✓	–	–	✓	–	✓	–	✓	✓	✓	✓	✓
Balance security with privacy and data protection	✓	✓	–	–	–	–	✓	–	–	–	✓	–
Institutionalise cooperation between public agencies	✓	–	–	✓	–	–	–	–	✓	✓	✓	✓
Foster R&D	✓	–	–	–	–	–	✓	✓	✓	–	✓	✓
Provide incentives for the private sector to invest in security measures	✓	–	–	–	–	–	–	–	–	–	–	–

Table A.36 Country codes
used in Table A.35

Code	English name
AT	Austria
BG	Bulgaria
HR	Croatia
CY	Cyprus
CZ	Czechia
DK	Denmark
EE	Estonia
FI	Finland
DE	Germany
HU	Hungary
IE	Ireland
IT	Italy
LV	Latvia
LT	Lithuania
LU	Luxembourg
MT	Malta
NL	Netherlands
PT	Portugal
RO	Romania
SK	Slovakia
SI	Slovenia
SE	Sweden
UK	United Kingdom

References

1. ENISA: National Cyber Security Strategies-Interactive Map. https://www.enisa.europa.eu/topics/national-cyber-security-strategies/ncss-map. Cited 1 Dec 2019
2. ENISA: National Cyber Security Strategies: Practical Guide on Development and Execution (2012). https://www.enisa.europa.eu/publications/national-cyber-security-strategies-an-implementation-guide/at_download/fullReport. Cited 1 Dec 2019
3. ENISA: National Cyber Security Strategies: Good Practical Guide (2016). https://www.enisa.europa.eu/publications/ncss-good-practice-guide/at_download/fullReport. Cited 1 Dec 2019
4. EUNITY Grant Agreement, Annex 1 Description of the action

Printed in the United States
by Baker & Taylor Publisher Services